TEACHING TECHNOLOGY

A How-To-Do-It Manual for Librarians

D. Scott Brandt

HOW-TO-DO-IT MANUALS FOR LIBRARIANS

NUMBER 115

NEAL-SCHUMAN PUBLISHERS, INC.
New York, London

Published by Neal-Schuman Publishers, Inc.
100 Varick Street
New York, NY 10013

Printed and bound in the United States of America.

Library of Congress Cataloging-in-Publication Data

Brandt, D. Scott.
Teaching technology : a how-to-do-it manual for librarians / D. Scott Brandt.
p. cm. — (How-to-do-it manuals for librarians ; no. 115)
ISBN 1-55570-426-3 (alk. paper)
1. Technology—Study and teaching. I. Title. II. How-to-do-it manuals for libraries ; no. 115.
T56.4 .B73 2002
607.1—dc21

2002002406

CONTENTS

LIST OF FIGURES

INTRODUCTION

GOALS

The goal of *Teaching Technology: A How-To-Do-It Manual for Librarians* is to introduce innovative educational concepts and techniques to help librarians and other information professionals create technology learning courses, sessions, workshops, and modules. Too often teaching and training put too little emphasis on the learner. No doubt you've heard the expression, "better to be a guide on the side, than a sage on the stage." This book, thoroughly grounded in the philosophy of instructional systems design (ISD), puts the learner first, not the teacher.

The theory of ISD grew out of the step-by-step approach used in the industrial trades to design useful staff training sessions. Soon the educational world adopted the winning ISD design to plan instructional modules. In addition to an emphasis on learner-centered instruction, Instructional System Design focuses on inventive ways to measure tangible results. How can instructors teach and train with the learner at the center of the learning experience? How can we prove that real learning is truly taking place?

Teaching Technology applies the ISD concepts to designing courses, or parts of courses, for instruction on information technology and information literacy. Essentially, this approach breaks learning down into specific pieces to create a highly structured—and successful—approach to creating learning modules, one-hour sessions and lesson plans. Traditionally, ISD deals with the interrelation of analysis, design, development, implementation and evaluation, a formula often referred to as "ADDIE."

I have adapted the ADDIE formula to provide a simple and effective approach to teaching for information professionals. My objective is to give not only structure, but also form and function, to the process of teaching library and information science.

ADDIE establishes five critical steps in effective learning design: analysis, design, development, implementation, and evaluation.

The ADDIE formula also emphasizes a dependent relationship between the steps. As Figure 1 demonstrates, analysis must be

Fig. 1: Diagram of ADDIE

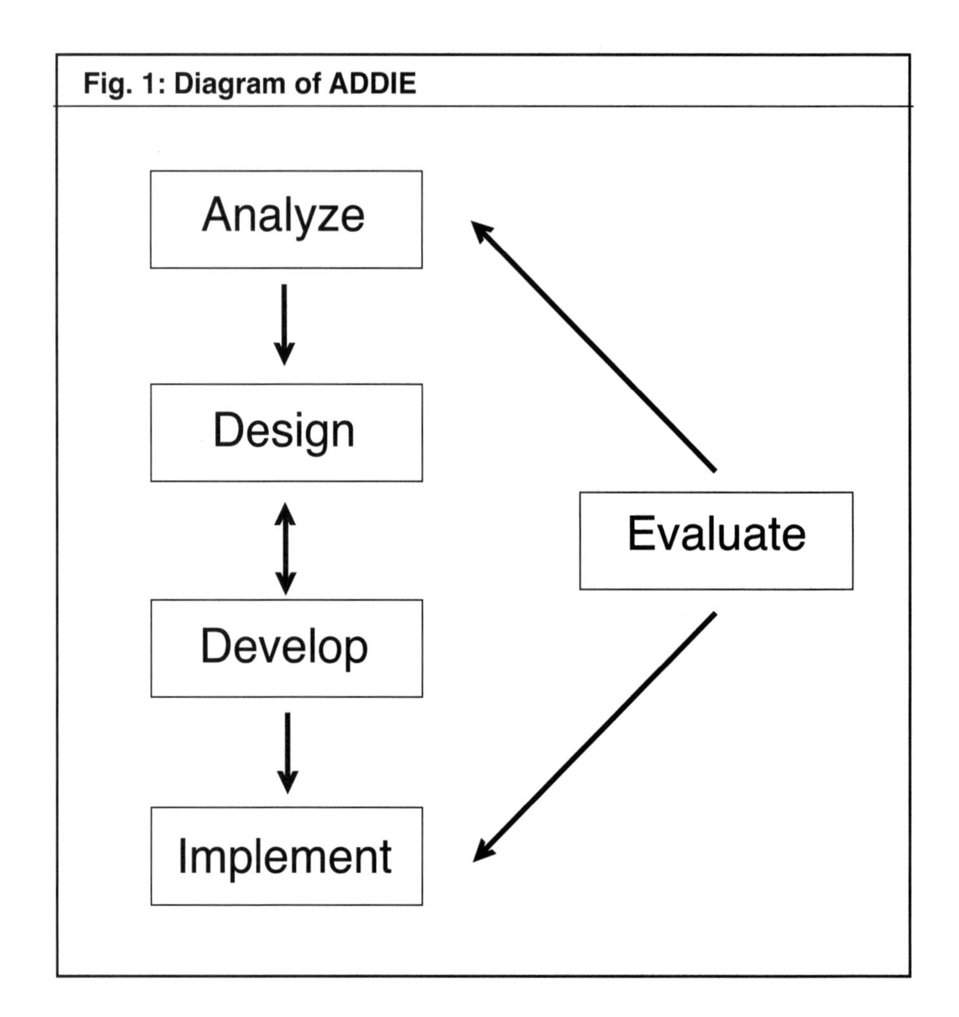

done before design, development builds directly off of design, and so on. The outcome of one step leads to the foundation for the next step. For instance, learner needs identified in the analysis step influence outcomes determined in the design step. This in turn influences objectives built in the development step.

This linkage creates a strong structure for the final product—the lesson plan. The plan will describe learners, outcomes, and objectives (and the steps to meet objectives) in a logical and practical way. The plan functions as a cohesive and effective facilitator of learning. In fact, carried out in complete detail, such a lesson plan should be able to stand alone as a self-instruction module.

STRUCTURE

Teaching Technology is divided into three parts. Part I, "Developing Technology Training Courses Using ADDIE," follows the formula step-by-step: it looks closely at analyzing learning, designing blueprints, developing instruction, implementing the teaching and eventually evaluating the results. A separate chapter is dedicated to each of the five steps, giving details on each one and describing how they inter-relate. Part II, "Building Effective Technology Training Programs," contains three chapters that discuss the overall program—how to build one, how to make it work, and how several actual programs have succeeded. Part III, "Sample Technology Training Materials from Successful Programs," offers a variety of useful, hands-on resources.

Each chapter starts with a set of objectives for learning. For example:

In this chapter you will learn to:

- Describe how to use a technology teaching program
- Identify ways to market and promote a program
- Discuss how a program can be implemented
- Define ways to make a program applicable
- Identify elements to evaluate a program

Teaching Technology guides you through the five main steps of systematically developing technology courses from scratch. Each step is a chapter that builds on the previous one.

CHAPTER BASICS: A BODY OF LEARNING DESIGN

Chapter 1, "Step 1. Analysis," is the brains behind the teaching of technology. This is a discussion of why and how learners and learning should be analyzed. It breaks analysis down into simple procedures, and gives instruction on doing four kinds of analysis. Chapter 1 also looks at how to analyze learners' needs—how to assess where they are now, where they should be, and how to bridge the gap. It addresses learner skill levels, attitudes, and learning styles. Finally, the five types of learning that effect the design and development of specific learning objectives are examined.

Chapter 2, "Step 2. Design," is the heart that pumps life into teaching technology. Here the designing process and the elements needed to create blueprints for technology courses are defined. This chapter demonstrates how to determine expectations, prerequisite learning, measurable outcomes and objectives from the learner analysis. It focuses on building practical four part learning objectives based on demonstrable outcomes. An explanation of the need for setting specific conditions and degrees when developing objectives ends the chapter.

Chapter 3, "Step 3. Development," is the hands that create the teaching technology product. It uses the blueprint and applies techniques to forge a complete course. How to integrate needs, outcomes, objectives and the steps required to create learning are covered. Chapter 3 also illustrates the process with specific concrete objectives. It culminates by discussing strategies to facilitate learning, and shows how to match appropriate strategies to different types of learning objectives.

Chapter 4, "Step 4. Implementation," is the legs that carry out teaching technology. This is where the aspects to be considered when you are face-to-face with students are addressed. The chapter starts out discussing specific attributes related to adult learners and their needs. It discusses various presentation formats and styles and outlines presentation approaches and successful techniques for presenting. There are also considerations here for using an assistant and the need for back-up plans. Finally, this chapter discusses how to handle problems with both technology and with participants.

Chapter 5, "Step 5. Evaluation," is the respiratory system of teaching technology, necessary to filter out toxins that might occur. It defines in detail different kinds of evaluation and why they are needed. This is the chapter that covers the evaluation of participants, the instructor and the course. Formative evaluation, which occurs during development, and summative evaluation, which occurs after presentation, are examined. Chapter 5 demonstrates how to create a number of different kinds of evaluations, and advises on when and where to use them. It describes various scales and survey instruments to collect feedback.

Chapter 6, "Building a Technology Training Program," is the embryonic beginning of teaching technology. It describes starting an initiative from the ground level. This chapter details what an initiative should cover and who should be involved in building it. The process of doing an "environmental review" to determine and prioritize training and teaching needs is reviewed. Costs and budgets are also addressed here, as well as the need to achieve buy-in and sign-off.

Chapter 7, "Making the Technology Teaching Program Work," is the sweat of teaching technology, illustrating how hard work pays off in the end. This chapter describes how to utilize a technology teaching program as a day-to-day operation to promote patron learning and staff development. It describes how to implement the program in a way to be applicable to your organization. The chapter ends with a discussion on how to market, promote and evaluate a program.

Chapter 8, "Exploring Successful Programs," is the smiling face of teaching technology. It is a collection of information from libraries around the country that describes aspects of their programs, including kinds of training/teaching offered, audience to whom training is offered, how often courses are offered, and what types of facilities are used. In addition, each librarian describes their most popular classes, successful techniques for teaching, and their proudest achievements.

Part III is "Sample Technology Training Materials for Successful Programs." This last section is selection of materials to demonstrate various aspects of the ADDIE process. One is an instructor's guide for an hour-long lecture/demo session on why indexes are better than search engines for finding information. Another is a set of handouts for a 90-minute lecture/demo session on a specific index (the EBSCOHost databases). Another is a set of handouts for a three-hour workshop on how to design and develop a module of instruction for teaching the Internet. Last is a syllabus and instructor's guide for teaching an eight weeks (16 hours) introduction to information literacy skills, specifically aimed at students in an engineering or technology program.

HOW TO USE THIS BOOK

Take your time, read through the chapters, then go back and try to build a module, looking back at the directions and examples. As you are reading, make notes. I encourage you to be passionate as you read—place yourself in the center of this learning experience! Take note of the things that jump out at you. Write down why it jumped out at you. Question what you see—does that work? Will it work in my setting? Where can I use this? What if I tried this combination?

Practice each of the five ADDIE steps. Find some learner aspect to *analyze*. Begin the *design* of a module. *Develop* some objectives and match them to strategies to create a lesson plan—in fact practice this several times. *Implement* it on somebody to test it out and get a feel for how it works in practice (even if it has to be your office mate or spouse). *Evaluate* something—either your own work or somebody else's. I enthusiastically invite you to become a dynamic part of the learning presented in *Teaching Technology* by interacting with the information wherever possible.

OUTCOMES

As you will discover, I use the word *outcome* (as do many others) to mean something very tangible. It is the physical result (either a thing or a demonstrable behavior) of learning something. If you learn how to catch a fly ball, for example, then the outcome will be a caught ball—not just the potential or ability to do so. In the example above, if you learn to identify specific things to do when marketing and promoting a program, you need to actually do marketing and promoting, and the outcome will be specific measures of success—more registrations, increased numbers of courses needed, etc.

After you have read *Teaching Technology* you should be able to create effective learning from start (analysis and design) to finish (implementation and evaluation). You need to actually analyze learners and the learning environment in gathering data and identifying trends. You need to design a blueprint for a learning module. An annotated outline or storyboard is what you are working toward. You need to develop objectives, strategies and materials. The outcome should be a complete lesson plan and the materials to support it. You need to practice the skills of implementation. Greater confidence in teaching and positive responses from students will be your reward. And you need to evaluate how you've done. Information to help improve the learning even further is the final outcome. The reasons why you need to do all that are explained in the chapters. But to prove that you've learned why, you need to actually do it and as a result have specific measurable outcomes.

You should be able to use some of the examples as a platform upon which to build. I've tried to provide a variety of things, from technology-based to library-based, from a variety of settings (university, public library, corporate). As an added bonus, I've included some descriptions of programs from those three areas in the hopes that they will give you insights and ideas, as well as challenge you to do the same or better!

ACKNOWLEDGMENTS

I have a lot of people to thank for their support, directly and indirectly, in beginning and finishing this book. Almost everyone who has come in contact with me—from my closest friends to colleagues with whom I almost never see but constantly e-mail—has in some way influenced me as a teacher. Linda Martinez and Eric Celeste at MIT were instrumental in getting me on the path of technology training, and I am appreciative of those early training sessions. The "conference gang," Jan Zastrow (Hawaii), Darlene Fichter (University of Saskatchewan, Canada), and Dennis Tucker (Northwestern State University, Louisiana), have been sounding boards and co-presenters who have given me feedback and ideas on various aspects touched upon in this book. Charles Harmon at Neal-Schuman has been a stern but understanding taskmaster and a very supportive friend from the inception (a process that he likened to a pregnancy, and at the eighth month I was begging for a cesarean). Michael Kelley from NS has been even sterner but very, very helpful—I don't think this book would have been as good without his guidance. Jane Dysart and Tom Hogan have always been very supportive in allowing me to do workshops at both Internet Librarian and Computers In Libraries conferences, which were the sources for a lot of the content of this book.

At Purdue University, I have to thank the Dean of the Libraries, Emily Mobley, who taught me the lesson of good management a long time ago when I ran out of student funds as Physics Librarian. Emily has always been interested in the directions my work takes me, and often offers suggestions, advice, or insight that has almost always been helpful. I'd also like to thank my boss, Bill Corya, Director for Information Technology, for being understanding and giving me the leeway to take on extra projects. Bill has offered support in the way of encouragement by telling me to keep at it because I'm doing a good job, or grounding me by asking "what the crap is that?" My co-workers in the Information Technology Department have always been good sports, willing to help with technical problems and putting up with my often zany antics. Thanks guys!

More than anyone, my cohorts in training have bolstered and challenged me. Kathleen Kielar, Staff Development & Training Coordinator, has introduced me to a lot of approaches and theories in education and training. She has shown me the benefits of organization and the rewards of analysis. She is wise beyond her years, though no one wants to admit it. Jennifer Sharkey, Technology Training Specialist, has been my equal since the day she started. I like to say that we are interchangeable in our responsibilities, but I am constantly reminded by staff who take her classes and the quality of the materials she develops that she surpasses me in several areas. She constantly increases her knowledge and skills, and is one of the most professional trainers I have ever met. I am lucky to have her as my training partner.

Lastly, I want to thank my dearest friends and family. Peter Janson, an outstanding professional musician and life-long friend has been there for me since the inception of the idea several years ago. His own dedication to his art, his creative outpouring, and his commitment to practice and performance have inspired me for a long time. He has chided me when I could use it, consoled me when I needed it, and has always made me think. Peter was actually my teacher, a long time ago, when I was in the Air Force. My affinity for training is in no small way due to his patience and thoroughness as an instructor.

My kids, Penny and Karl, have always been a source of inspiration and joy. Both musicians (they get it from mother's side), they are creative, exuberant, and always thirsty to learn more. I like to think that I instilled that sense of a quest for knowledge in them, and it touches and inspires me to see them pursue their dreams. In a way, they were my first "guinea pigs" as students, if you don't count my attempts at age 10 to moderate a game of *Jeopardy* with my siblings (I'll take Art Fleming over Alex Trebek any day)! From the time they were born until they reached adulthood they have been a constant source of inspiration for learner-centered teaching and parenting.

Finally, I would like to thank Nancy Clement, my partner, confidante, and source of contentment. Together we have kissed the Blarney Stone, tripped the light fantastic in New York City and London, basked on the beaches of the Virgin Islands, and danced to the blues in Chicago. Through it all she has had to put up with my frets and worries, my confusion and frustration, my triumphs and bragging. I have bounced a lot of ideas off her, some that have soared while others have fallen flat. She has endured with me all the manifestations of stress of completing this book, from grumpiness to thinning hair. And she has put up with all the long hours I have locked myself away to write. I owe her a debt of gratitude for her support, understanding, and care. Thank you, Ginger.

PART I

DEVELOPING TECHNOLOGY TRAINING COURSES USING ADDIE

1 STEP 1: ANALYSIS

In this chapter you will:

- **Describe importance of analyzing learning aspects**
- **Define learner analysis**
- **Describe three reasons for analyzing learners**
- **Identify four categories of analysis**
- **Determine learners' needs**
- **Describe importance of identifying learners' level of experience**
- **Identify four aspects of learner attitudes**
- **Define five types of learning**
- **Discern between styles of learning**
- **Review template for performing learner analysis**

INTRODUCTION

Analysis. The very word can send a shiver down the back or tie a knot in one's stomach. It can evoke feelings of math anxiety or statistics-phobia. Analysis may be a dirty word to some, or at least conjure up nightmares of mathematical calculations and using formulas with names like *chi squared* and *standard deviation* to determine seemingly obscure results. However, in its broader sense, analysis means simply to analyze, or review with scrutiny. And that's what is meant here—to scrutinize or look closely at learning before designing and developing it.

Analysis can be thought of simply as "thinking before doing." How often have we all been guilty of jumping right in and starting something, whether teaching or anything else in life, without analyzing the situation first? When we stop and think, we take the time to consider aspects that we glossed over or didn't even notice. And since learners are a very important part of learning, we want to stop and give some thought to them as part of the learning equation before we go any further.

This chapter discusses what it means to scrutinize various learner elements that can and should be analyzed prior to developing learning. It emphasizes why learners need to be analyzed and gives examples for doing so. It is hoped that it will reveal or reinforce the need to consider the learner before, not just during or after, the learning. It addresses aspects of specific learners (needs, experience, attitudes) as well as aspects of learners in general (learning types and styles).

IMPORTANCE OF ANALYZING LEARNING ASPECTS

Given that you have a direction for your teaching, the first area that needs consideration is the learners who will be receiving it. Considering your learners means gaining insight, understanding, or sympathy related to their needs for acquiring knowledge or skills.

One of the critical things for you to understand as a teacher or trainer is that students bring attitudes and aptitudes to a learning situation. That includes their learning styles, preferences to teaching, and mental models of understanding how various things work. The more you understand them, the easier it will be for you to help facilitate their acquisition of knowledge.

This is a part of teaching that is often overlooked. Some possible reasons for ignoring this area include:

- An attitude that you are the keeper of knowledge and since you are going to share it with others the least they can do is just sit there, acknowledge you, and accept it
- A lack of time to carry out any survey or questioning
- Fear of what you might find out interacting with learners
- Misunderstanding of how it can benefit you

If you have the attitude that learning is a one-way street and learners should absorb (pay attention to, write down, memorize) your every word, you will overlook a key element in learning—the learner. How are you going to account for her? If you don't have time to spare to analyze prior to designing, you are going to miss out on a crucial perspective—the learner's. How are you going to make time to inquire and listen? If you are afraid to find out what learners really want, you're really teaching for yourself, not for them.

What happens when you simply try to "throw" knowledge at learners and expect them to absorb (and appreciate) it? It often bounces off. Current theory in education argues that learners need some "buy-in" to acquire knowledge. What would have been called "catering to them" in the 1950s is considered "smart sci-

Fig. 1.1: Comparison of Teaching

Old fashioned teaching	ADDIE-influenced teaching
"thrown into it" and "hope it sinks in"	"buy-in" approach
memorization	learning by understanding
passive, quiet classroom	interactive, exciting classroom
aimed at children as students	aimed at learners of all ages
focuses on similarity of student needs	recognizes & celebrates indiv. differences
focuses on rules of instructor	focuses on motivation of students
gathers statistics for analysis	determine strengths

ence" in the new millennium. Understanding where their attitudes and motivation come from is savvy teaching.

This is different than what some may call an "old-fashioned" approach to teaching. My father used to speak of teachers in the 1930s and 40s who ruled by the "iron fist" and believed the old adage "spare the rod, spoil the child." Back then, it was more important to control behavior in the classroom than to inquire about the person attached to the behavior (see Figure 1.1). It was also believed that students learned best by sitting perfectly still and memorizing facts, figures, and formulas. Under those circumstances, it didn't seem important whether some students learned by doing instead of memorizing, or by reading rather than listening.

Let's face it, the majority of teaching and training we're talking about deals with adult learners, not children in their early stages of development. Adults have different needs and different attitudes to learning. They are better at assessing their situations and identifying their needs. And they are better at articulating them.

Of the many different students I've seen in classes, workshops, and online, it always amazes me that no two are the same. I've taught to high school students in gifted classes, senior citizens in a lifelong learning center, college students in a required information course, community leaders in a small town, and information professionals in national and international settings. But interestingly, there are many similarities between all of those groups of learners. They must have motivation for learning, they all bring certain experiences with them, and many of them have similar learning preferences or styles. It is also amazing that it is so easy to misinterpret their confusion for boredom, anxiety for anger, and concentration for complacency. Problems related to, or the lack of motivation on the student's part is covered further in chapter 4.

Fig. 1.2: Example of Gathering Analysis

Respondent	Info searched for	Currently use	Would like to	Want training?
1. Senior	Genealogy	Netscape search	Be comprehensive	Maybe
2. K12	Game shortcuts	Yahoo	Search faster	Not really
3. College	Term paper	Nothing	Find answers	Real quick
2. Senior	Travel info	AOL	Find answers	Mornings only

Some people think that if you gather data, you've done analysis. Unfortunately, that's not true. First you have to decide what to gather, how to gather it, and then do the actual gathering. If you've created a spreadsheet or database, you've probably spent some time thinking through how to collect and organize data. Once you have your data, then you can begin looking at it to determine deficiencies, patterns, or trends.

An example of analyzing: A simple analysis can start with a survey, as described in more detail below. Let's say we have a group of public library patrons whom we suspect want (and perhaps need) training on how to use Internet search engines. You could ask a few people if they would like the training, and then determine whether the Aye's outnumber the Nay's (or vice versa). However, it is important to keep in mind two things. First, the more different kinds of people you ask, the more reliable your overall picture will be. Asking a group of high school kids will likely yield a different result than asking a group of seniors. Second, the more detail you can ask, the more depth your analysis will have. Rather than simply ask if they would like search engine training, you might ask what types of things people search for, where they search for them, what kind of outcomes they achieve, how they measure success, etc. An "analysis data collection tool" might look something like that shown in Figure 1.2.

Why analyzing is important: You know what they say about assumptions...and guessing can get you into just as much trouble. Analysis is important because by using it you can get more accurate insight into designing learning. It helps us form a more complete and objective assessment of the circumstances involved. It allows us to make better and more informed decisions.

Fig. 1.3: Key Questions For Learner Analysis
Key Questions Checklist for Learner Analysis
√ How do I determine what patrons/learners want? √ What would a successful outcome look like? √ How should I interview and ask questions? √ How do I determine the best sample? √ How do I analyze results? √ How do I translate results into a new design for learning?

DEFINITION OF LEARNER ANALYSIS

A general definition of learner analysis is: the inspection of attributes that will impact the learner who is acquiring knowledge. Learner attributes are their knowledge/skill needs, current depth of knowledge/skill, and their attitudes and approaches to learning. Simply stated, to analyze learners one must investigate who the learners are, what they need, and how they feel about it (see Figure 1.3).

How do you analyze learners? Analysis is really a part of a process of determining the extent of a situation, issue, or problem and looking at it closely. The steps for doing learner analysis are:

1. Determine basic area of need (What does it seem like they want?)
2. Identify likely outcome (How would you describe success?)
3. Determine questions to ask and method for asking them
4. Interview as many people as is reasonable
5. Analyze results, looking for trends, discrepancies, gaps
6. Use the resulting analysis to help design learning

DETERMINE BASIC NEED

The need for learning stems from some need for improvement, often recognized by the learner as a gap in skills or knowledge, but sometimes mandated by changes in a work environment or the introduction of new technology. A general sense of that need usually arises in the form of problems, confusion, agitation, or even disgust. For instance, if more than one library patron screams, "I can't find anything on the Internet!" you get the general idea that there might be a need for learning. Thus you could generalize this basic need as "help finding information on the Internet."

IDENTIFY LIKELY OUTCOME

Given a basic need, can you visualize what a successful outcome for satisfying the need might look like? Specifically, an outcome should be a demonstrable or measurable result, behavior, or change. The outcome might be a satisfied student or patron, of course, and/or useful information, obviously. But what else? Perhaps the outcome might include knowing where to go, which tools to use, or how to use them to search efficiently. First try to identify what the outcomes might be, then start thinking about goals to meet them.

DETERMINE QUESTIONS AND METHOD FOR ASKING

Ask yourself questions about how you can measure something to evaluate it. For instance, if you think patrons use one search engine over another, how can you measure that? You could, 1) ask patrons throughout the library; 2) interrupt patrons using computers and ask them; 3) randomly check what is on the computer screen through unobtrusive observation; or 4) periodically review the browser log to see which search engine URLs appear.

By determining specific outcomes you can shape the questions you ask so that you elicit specific responses. Too often when vague or general questions are asked, vague or general answers are given! Rather than ask, "Do you need help searching?" you might ask, "What kinds of things are you trying to find?" or "What might help ease the frustration you currently have?" Likewise, when you are gathering other data, be specific about what it is you want to capture. Are you interested in which search engines they use, or whether they use a directory-based versus non-directory? When you can elicit specific responses you will analyze learners' situations and needs better.

Once you have an idea of what you want to ask or look for, you need some kind of methodology for asking and keeping track of the responses. The section below on categories of analysis addresses general methods for gathering data. Basically you have to determine whom you will ask, when, and where, in addition to how.

INTERVIEW PEOPLE

Determine what your "market population" is, and ask away. You can either do a random sample, asking several people to get an "average" response. Or you can do a representational sample, making sure to ask representatives of the population (e.g., if in-

Fig. 1.4: Key Questions For Analyzing Results
Key Questions Checklist for Analyzing Results
√ Do I have all of the results in one place/format?
√ What patterns or repeat issues/problems can I see?
√ Are some things mentioned more/less than others?
√ How do I translate patterns or trends into needs?
√ With whom can I verify/double-check needs?

terviewing public library patrons, you might ask youths, teens, adults, and seniors; if interviewing within a library work environment you might ask reference librarians, circulation assistants, and catalogers). Realistically, not everyone you ask will want to or be able to participate, so you must decide how much of a sample will be suitable.

ANALYZE RESULTS

As noted, analysis means to scrutinize. Look closely at the responses given to you (see Figure 1.4). Are there any patterns or trends that emerge? Perhaps one frustration in particular stands out, for example retrieving too many results when using *Yahoo!*. Did the responses verify or validate the need you identified and the subsequent outcomes? For example, you might determine you need to explain how to use *Yahoo!* rather than try to switch them to your favorite search engine. Are the responses very specific and varied? For instance, if 10 out of 20 people mentioned using two to three different search engines, maybe you'll have to consider demonstrating more than one search engine as part of your training. Sift through the wealth of data you gather and try to identify what might help the learners best.

DESIGN LEARNING

Once you have gathered input and given some thought to what it means, you're ready to move on... to the next chapter, that is. Sometimes the relationship between analysis and design will seem like one of those "which-came-first-the-chicken-or-the-egg" kinds of things. To identify some of the things that you want to analyze, you might have to have a design in mind. And there will be times while designing that you think of something else to analyze. There is no strict sequential order—it's okay to do a little of one while you're doing the other.

An example of learner analysis: The example above about searching on the Internet is part of a process of analyzing learners. The whole process might include: asking staff in the library what kinds of complaints they have heard about searching on the Internet; formulating your survey; testing out some questions on a patron to see if you are expressing things in a way she will understand; checking in with the director (or library board) to find out if there is a preference in how do to the survey; having settled on a quick "exit interview" format, creating specific questions; conducting the survey over a sufficient period of time; organizing the data you collected (perhaps in a table format); sitting down with a colleague to scrutinize the data; and checking your conclusions with management prior to making a final proposal for learning. Then you can start designing.

Why learner analysis is important: Not to beat a dead horse, but unless you really know what the learner wants or needs, you're merely spouting knowledge and not helping them to learn. We all like to think we know, or can spot an obvious need, but it is important to include the learner in the process. They can give the best insight into what would be helpful and useful. Besides, not only might you be surprised at what pops up, the learners will appreciate that you asked them.

REASONS FOR ANALYZING LEARNERS

I'll admit that I've been guilty once or twice of not analyzing my learners. When this happens, there is a chance that the teaching will encounter some difficulties. Foremost, students could get bored and "turn off" the learning. Or it might be hard for them to relate to the level or amount of knowledge or skills being taught. In addition, learners can become resentful, confused, or simply walk out if the learning doesn't meet their needs.

As noted, learners must have a motivation for learning. They bring with them differing experiences that may be applicable to the learning situation. And they have various learning preferences or styles that will likely affect how they react to the learning.

One of the most important reasons to analyze learners is to find out about their current level of knowledge about the subject being taught. What do they know now? What do they want to learn? What might they learn to do differently to help them acquire knowledge or skills? Another reason is to find out about the learners' styles and preferences for learning. Is there a way to hone in on the way they learn to help facilitate the learning? Additionally, it is useful to investigate learner attitudes toward the subject (or learning in general). What is it like from their perspective—how willing are they to undertake the learning? What, if any, misconceptions about the topic do they have? And where or when will they applying learning?

To sum up, the reasons for analyzing learners are:

- Ensure that the learner isn't ignored or forgotten when designing learning
- Identify the learner's needs or gaps in knowledge or skill to ensure learning is aimed at the right target
- Determine the learner's current level or related expertise to help judge how to address learners during the learning
- Objectify learner attitudes towards the learning to motivate them to participate and cooperate fully to their satisfaction

CATEGORIES OF ANALYSIS

At this point you might be thinking, "Okay, but *how* do I do this?" The best answer is probably "thoughtfully and patiently," but you'll probably want something a little more concrete than that. However, remember that thinking carefully as you work your way through the process is important. And you do have to be patient to proceed—some people don't bother because they want to avoid the work and aggravation.

To make analysis simple, it is useful to divide the approaches to analysis into four categories: direct, indirect, formal, and informal. Direct analysis requires interaction with the potential learners in order to hear their needs. Indirect analysis implies interacting with others, or through some means of observing learners without interacting with them. Formal analysis means gathering data and recording specific and precise responses from learners. Informal analysis implies collecting or validating general impressions of the learner.

As shown in the grid in Figure 1.5, it is helpful to consider combinations of direct/indirect and formal/informal analysis. Each combination has a slightly different approach and use.

A direct-formal analysis requires the most work and time in defining, preparing, gathering, and reviewing information. It can be difficult when the person who is being interviewed finds it hard to be articulate or objective about what she or he needs, which means the questions should elicit precise answers. Thus, it usu-

Fig. 1.5: Matrix of Direct/Indirect and Formal/Informal Analysis

	Formal	Informal
Direct	Survey/ Anonymous form	Conversation/ Short e-mail
Indirect	Workplace interview	Observation/ past experience

ally provides the opportunity to ask a lot of specific questions, which makes it worthwhile. Anonymity, which can be accomplished via a Web form, might encourage participants who are otherwise reluctant to speak up. By the way, formal surveys or questionnaires usually take time not only to create but also to get approved by "the management." Often the information must be shared for reasons other than designing the learning, so be prepared to distribute your findings.

An indirect-formal analysis can be almost as time-consuming as the direct one. However, because you are interviewing someone other than the learner, the information you get may be more objective. For instance, a supervisor is likely to know and be able to articulate the expectations placed on a job or staff member. A teacher is probably more likely to be able to describe what a student would need from certain training. Obviously, there are situations where this kind of analysis may not work because there are no external expectations on the learner. For instance, independent learners such as adults are likely to be internally motivated and there is no one to whom they are held accountable or who could be interviewed.

A direct-informal analysis gives quick and general, although usually very useful, information. It is likely to be less intensive and gather less information than a formal one. The length and number of questions will be less than with a formal one. But the trade-off is that you might be able to ask more people a smaller number of questions. The informal or perhaps "conversational" format lends itself to more personal interaction and follow-up. It is possible to ask the questions without alerting the participant to the intent of the questioning, which has its pluses and minuses.

An indirect-informal analysis is probably the most vague way to gather information, but is (unfortunately) possibly the most often employed. Think about it: you're not really talking to the people who will be participating, and you're going about it in a way that is subtle and somewhat unreliable. It usually involves subjective interpretation on the part of the analyzer, and thus shouldn't be entirely trusted. Not that you can't rely on your own observations, or inferences made by overhearing complaints. But indirect-informal analysis is better than nothing because, if you are in a situation where you are prevented for some reason from dealing directly with the learners beforehand, you'll have to somehow make do. Try, however, to take on the persona of the learner when you're making guesses—if you were the learner, what would you want from the learning?

It is possible, and encouraged, for teachers and trainers to combine these categories of analysis. In fact, it is not always possible to use one exclusively due to time and population sampling limitations. You may have a small window of time before you have to begin the learning, or interviewees may be willing to give only small amounts of time. The sample of your population may be limited for a number of reasons—some people may be unapproachable, others may be unwilling to participate. In some cases you may be able to do only a few extensive surveys and many short ones. In other situations, where you have indirect information, you may want to verify or validate your findings with some direct input.

An example of direct-formal analysis method: Let's say you perceive problems amongst your staff using e-mail—many complain about being overwhelmed with reading, responding to, and keeping track of it. You realize that because more and more of your services involve e-mail, staff should be highly organized, but it doesn't appear that they are. So, your outcome or hypothesis is that staff are not as skilled at managing e-mail as they can be. How do you analyze this? You have to identify those things that you rate as exemplars of good e-mail management and survey the staff to verify whether or not they use such skills.

You might ask an expert what it is that qualifies as good e-mail management skills. Or you might try to identify those skills yourself, either by reading through e-mail training guides or asking savvy users what their "best practices" are. What you need is a benchmark of what it means to be highly skilled. As you gather this information you will be gathering the information for questions you will ask of staff. For instance, if someone says, "I quickly scan and rate my e-mail as 'rush, important/must read, interesting and delete-able' first thing in the morning," you have identified a question: "Do you assign a ranking and sort e-mail first thing?" Or if someone says, "I put informational e-mails into a folder and save them for the afternoon," you may have identified another question: "Do you use folders to separate mail according to type of subject?"

Fig. 1.6: Survey Questions

Survey on how you currently deal with e-mail

Circle a number to answer the following questions…	Never	Not Often	Not Sure	Often	Always
Do you check your e-mail more than once a day?	**1**	**2**	**3**	**4**	**5**
Do you prioritize e-mail in your in-box?	**1**	**2**	**3**	**4**	**5**
Do you use folders to separate e-mail?	**1**	**2**	**3**	**4**	**5**

Fig. 1.7: Sample Analysis Spreadsheet

Survey on the use of e-mail features and e-mail management

Name	Lib Unit	Checks e-mail	Sends	Prioritizes	Deletes	Use Folders	Attachments
Adams	TS	4	2	2	4	1	4
Campbell	PS	5	4	4	5	2	5
Delgado	IT	5	5	4	5	4	4
Ford	TS	4	4	2	3	3	2
Johns	PS	2	2	1	1	1	2
Klingmore	TS	4	4	3	2	1	4
Schott	TS	4	4	1	4	1	4
Turner	IT	5	5	2	2	3	2
Whitcombe	PS	2	2	3	1	1	1

Eventually you will end up with questions on a survey that may look like that shown in Figure 1.6.

Note that the measurement scale in this survey requires questions to be asked in a way that elicits responses that fit the measurement. In other words, you can't ask, "Do you check e-mail?" Instead you have to ask, "Do you check your e-mail more than once a day?" and then allow them to respond, "Never. Not Often. Not Sure. Often. Always." You may need to include a different scale of measurement, or "fill-in-the-blank" type of questions, such as "How often do you check e-mail? Hourly? Twice a day? Daily? Twice a week? Weekly?"

Such a survey could be a printed form, e-mail message, or online Web form. The more responses, and the more variations in types of responder, the better. Note that the form above lends itself nicely to creating tables or spreadsheets to tally and record the data. Once gathered you will likely need to summarize, for your own analysis and possibly to inform others. By looking for patterns and trends, you should be able to get a good idea of what the needs are, and, in turn, what the learning should cover (see Figure 1.7).

Why using different categories of analysis is important: As noted, some analysis is better than none, but multiple analyses are best of all. The direct-formal survey above may actually be preceded by a direct-informal (asking staff casual questions) survey or indirect-formal survey (asking supervisors about learning issues) to identify what should go on the staff survey. Remember that each category has its place: you might be able to get permission or cooperation for one more than another. And if you get the feeling that analysis sounds like asking questions and gathering information all the time, you're getting the hang of it!

DETERMINING LEARNERS' NEEDS

First and foremost, you need to analyze what it is the learners need. As noted above, you can ask in a variety of ways, but the bottom line is that a learning need has to be present. There are times when the need is not a learning need! It may be a lack of hardware or software, or it may be a matter of workflow and task distribution. Sometimes there are personnel issues that seem like learning problems, but reveal the need for changes by the management, not training. For instance, staff might refuse to use a software program because they feel it is a threat to job security—if the program allows one person to do the job of three, someone might get fired. Needless to say, no amount of training in the world can remedy that situation.

If you're lucky, someone will identify the need for you: staff need to know how to create word-processing form letters; patrons need to find genealogy resources; students need to be able to correctly cite electronic resources. You might simply identify certain training on the basis of new product arrivals or upgrades to new versions. Otherwise you might simply be told, "We need training, create a program," or, "We just got funding for training, what should we do?" If your job is training you need to stay on top of what staff and patrons are doing.

In lieu of having training or teaching needs handed to you on a platter, or even if they are, you need to assess situations and investigate likely outcomes. The best place to do that is in the environment where the potential learners are working and encountering problems. Visit staff on the job. Talk to patrons or students at the computer while they are searching. Ask to stop in at the businesswoman's office to see what goes on.

Simply stated, learner needs can be expressed as a gap in knowledge or skills. Identify where the learners are currently, where they would like or need to be, and determine what it would take to get their knowledge and skills updated. If learners profess to having a hard time searching on the Internet, determine what they would like to be able to do (work faster, have fewer search results to scan) and compare that to what they know or can do now. Once you've identified those two perspectives, figuring out how to bridge the gap is much easier. As they say, defining the problem is 90 percent of solving it.

It may be possible to combine the "where-they-are-now" with the "where-do-they-want/need-to-be" in one survey (probably a long, formal one). Or it might be necessary to use a combination of the direct-indirect, formal-informal methods. In any case, be sure to ask specific questions. Rather than ask, "Are you satisfied with searching?" ask questions about what they are trying to achieve, "Do you get too many unrelated results when you search? Do you regularly use more than one search engine? Do you mostly do one-word searches or do you use phrases?"

An example of determining learners' needs: You may be approached by someone else to teach or do training on a particular topic. However, you must still do some analysis. As noted in the example above, library management could ask that you do staff training on managing e-mail. You will need to ascertain that it is a learning issue, apart from problems with resources, time, etc. Check with front-line users to find out what they are doing, what problems they are encountering. A direct informal approach is probably best. Talk to both staff and their supervisors. Ask specific questions about all aspects related to e-mail—composing, sending, receiving, dealing with attachments, managing messages, and using folders. Where possible, find out *why* they do what they do currently to gain insight into what might be preventing them from doing things differently. For instance, maybe they don't use folders because they are unfamiliar or uncomfortable with folder hierarchies in general, a not uncommon occurrence. Or maybe they don't understand that they are allowed to reorganize the in-box, and thus haven't tried any options for doing so. Once you have a good feel for the issues involved, consider doing a larger survey to find out the extent and scope of the problem. You might find out that one group is effectively sorting e-mail, whereas only a small group of people you interviewed have needs for training.

Why determining learners' needs is important: As the example points out, you don't know what's needed until you investigate. And you're not just asking questions, you're pulling together pieces of a puzzle, or making a map. You have to find out what the present holds (issues, problems, etc.) and what the future could bring (aspirations, outcomes, etc.). You need to find the pieces that will complete the picture and bridge the gap, and you do it by talking to or observing the learners. Remember, the learning is about their needs, not yours.

IDENTIFYING LEARNERS' LEVELS OF EXPERIENCE

Determining needs is a specific analysis—it assumes that needs are particular even though they stretch across a large population (e.g., everyone needs to learn how to use a new version of software). Identifying level of experience is more of a general analysis—it seeks to gauge the depth of experience or ability of a variety of learners so that you can make some generalizations (i.e., "How many people are novices to using spreadsheets compared to those that have extensive experience?"). You need to know what variations there are among the learners who will be engaging in the learning. In other words, yes, everyone may use e-mail, but have some of the learners been using it longer, or do some of them have an understanding of folder hierarchies from previous applications?

If you haven't already dealt with this issue, it becomes very important when trying to facilitate in-class learning with a group that has varied experience. It is quite possible that some of the learners may get bored because they already know the things you are trying to teach, or that others will get frustrated because you're not giving them enough background and it's all going over their heads. Once you find out what they need, you'll need to find out how much they already know—the extent or depth of their knowledge and skill.

The way to do this is to include questions on a survey that get at the level of the learners' knowledge applicable or related to skills. For instance, if someone has used several search engines, she probably has a better conceptual understanding of what she is doing than someone who has only used one. The prolific searcher may still need help with searching, but the level or type of learning is going to be different than with someone who is more of a novice.

You need to gather information and analyze. Are there differences within the levels of knowledge or skill of the learners which impacts on their needs? So, in addition to asking what their needs are, ask about their background, training, etc. Be sure to ask questions in a way that does not put the learner ill at ease. Try not to ask questions that learners may feel embarrassed to answer honestly—for instance, if there is only a "Yes/No" response and answering "No" would make their skills seem less than adequate. Perhaps rather than saying, "Can you do *X*?", you might ask, "How often do you use *X*? Never, rarely, not sure, sometimes, frequently, or all the time." Ask how comfortable they feel on a scale of one to ten.

The ideal thing would be to screen all participants to check their levels. This is possible, to some degree, if you have a structured registration program in place. When learners sign up you ask them questions about their skill, formally or informally. This could be a short form with questions pertinent to the specific learning for which they are registering. They could be asked to prove or demonstrate the requisite knowledge or skills through a test (a short online Web-based quiz, perhaps). For online learning systems you can direct learners to start at different points, based on their level as determined by a survey or quiz.

But what do you do when you don't have registration? One suggestion is to use a direct-informal survey "on-the-spot" as learners show up—easier to do in small "face-to-face" training sessions than large lectures or online situations. Another is to take a moment to first emphasize the objectives and prerequisites for the learning. If it is an introductory class, you might warn advanced users that some of the material will be familiar to them. If it is an advanced class, you might warn beginners that the material will be confusing to them. In a classroom you will soon get an idea of levels by the look on the learners' faces or their body language. In some situations the learners won't have much choice over participation, for instance in mandatory classes. Don't panic if some start to leave, or act like they wish they could—you can only please "all the people all the time" in a perfect setting.

What do you do with the information you collect? There are four schools of thought on dealing with levels of experience, knowledge, and skill, especially in hands-on sessions (see Figure 1.8). The first approach is to segregate learners. Divide between advanced, intermediate, and beginner (and sometimes remedial). Unfortunately, this is usually only possible in some highly structured training programs. A second approach is to ask those with more experience to sit with someone who has less experience and help out. This approach seems to get mixed reviews—some people hate being put on the spot, while others love to help out. Ask for volunteers before doing this, rather than making it mandatory.

Fig. 1.8: Dealing With Various Levels of Learner Experience

Four approaches to varying levels of learner experience or skill	
Segregate: Divide courses between novice and advanced levels of knowledge/skills	**Collaborate**: Ask those with greater experience to sit with and guide novices
Roving Assistant: Employ a co-trainer to deal with questions that come up	**Suppl. Material**: Supply background info for novices, detailed info for advanced

The third approach is to include a "roving assistant" to help with learners who have lower skill levels, as they are the ones most likely to slow down or interrupt a learning session. Sometimes the "rover" is also able to answer sophisticated questions that advanced learners have. A fourth approach is to include supplementary materials for both introductory and advanced learners—simple conceptual explanations for beginners, and more detailed examples for advanced learners. This is something of a compromise that can work in lectures or online systems.

As we will see when talking about objectives, one way to address the problem of level of experience is to let the learners assess themselves. State plainly (and firmly) the prerequisite knowledge or skills that are necessary prior to undertaking the learning and the objectives that will be covered. For instance, publicize that an introductory class must be taken (or that mastery over "mousing and keyboarding" is required) prior to class that covers how to use a Web browser. However, because some people over-assess themselves, they may believe their experience (two years' worth) qualifies as extensive, even though in reality their skill is low (trouble double-clicking and highlighting). What to do about this is covered in more detail under implementing learning.

An example of dealing with learners' level of experience: Let's say that as part of an informal analysis for a course on managing e-mail you notice a lot of people with hundreds of messages in their in-boxes. You discover that using folders might be one aspect of the learning, so you want to find out to what extent people use them now in any capacity: e-mail, word processing, general file management, etc. It may be that people know how to use folders, but haven't used their knowledge and skill in this application. (If that is the case, it may be a management matter rather than a training one.)

Thus, questions you ask would be centered on folder use and experiences in general, not solely in the e-mail application. The options for responding shown in Figure 1.9 take into account

Fig. 1.9: Sample Feedback Survey

Survey on Computer Skills Applied to E-mail	
How comfortable do you feel accessing folders on the hard drive?	1) Try not to use them at all 2) Feel lucky when I find the right one 3) Okay, but sometimes it takes too much time 4) Successfully organize my files in them 5) Understand hierarchical directory structure
Have you used folders to group e-mail messages?	1) Didn't realize you could 2) Thought about it, but never tried 3) Tried to, but couldn't figure it out 4) Did once or twice, but didn't like it 5) Yes, I use them now

several possible levels of expertise, and are worded in a way to make answering easy for the learners. (Note: If you have doubt about the wording, ask a typical learner for feedback when developing the survey.)

Why dealing with learning level of experience is important: It is critical to gauge the learners' previous knowledge and skills because they are the recipients of the learning—to teach them you need to know where to begin and how much or how little can be taught. As noted, learners may get bored if they already have the knowledge being taught, or frustrated if the learning is beyond their capacity when they lack prerequisite skills. A bored learner not only "turns off" but may also become a distraction. A frustrated learner is likely to become angry or accidentally disruptive by asking a lot of questions.

LEARNER ATTITUDES

Separate from the affective or attitude type of learning noted above, learners bring attitudes regarding learning in general to a situation. Teachers and trainers should take into account various aspects of attitude that can shape the learning. Are the learners there because they are required, or because of self-interest? Have they had recent good or bad experiences with similar training? Do they have motivation for undertaking the learning?

A popular approach for dealing with the attitude and motivation of learners toward learning is known as the ARCS model.

Developed by John Keller, each letter represents an aspect that the teacher or trainer should include to insure successful learning: Attention, Relevance, Confidence, and Satisfaction (Keller, 1983). Each aspect should be addressed to increase motivation and thus address concerns that impact attitude.

ATTENTION

How can you get the learners' attention in the first place? What will catch their interest? Sometimes this may include the use of a gimmick, such as handing out candy, but ideally would relate to the learning at hand. Perhaps some surprising facts could be conveyed which relate to outcome of the learning ("Sorting e-mail into folders saves ten minutes per day!"). Or a declaration of skills to be learned could be emphasized ("By the end of this class you'll be able to search nine different search engines!"). Maybe learners would be told that for their participation they would walk away with an interactive tutorial on a disk. In an information literacy course instructors used a role-playing exercise involving solving a problem before a simulated bomb went off—the game introduced various concepts related to group work and organization of information.

The bomb exercise, developed by the author and Alexius Macklin at Purdue University Libraries, served as an icebreaker and attention getter as part of a problem-based learning approach. A mysterious package with a timer counting down was brought into the room after attendance had been taken. The students were told that they had a small amount of time to correctly identify the six elements of the bomb (power source, switch, timer, ignition, detonator, and explosive) before the timer went off. The students were then given packets of information that included numerous pages from a variety of sources. For instance, the group of students who had to find the correct timer sorted through pictures of clocks, diagrams and schematics, an encyclopedia article about an artificial cow heart, advertisements for watches, etc. The point of the exercise was to demonstrate how hard it is to find and evaluate information.

RELEVANCE

You need to understand what learners feel is relevant and include examples, exercises, and strategies to meet their expectations. If your learners are attorneys or physicians, they would likely prefer searching examples related to law or medicine as opposed to pop artists or body piercing. Think about the background of your learners and use appropriate examples (see Figure 1.10). If you aren't sure, ask! Relate student learning to their classes, as well

Fig. 1.10: Sample Topics For Various Audiences

Group...	Possible Relevant Topics...
High School Students	Colleges, Music, Exotic Lands & Animals
College Students	Jobs, Learning Aids, Consumer Technologies
Seniors/Adult Learners	Investing, Travel, Genealogy, History, Gardening
Not For Profits	Sources of Funding, Free Services & Products
Teachers	Learning Skills, Lesson Plans, Subject Content Sites
Chamber of Commerce	Economic Forecasts, Demographics & Statistics
Physicians	Medical Alerts & Trends, Specific Research
Attorneys	Business/Law News, Specific Research
Librarians	Internet Search Engines, Digital Technologies
Non-professional	Genealogy, History, Online Buying, E-mail

as topics that interest them. Relate staff training to work-based topics. Relate learning for others to topics that are relevant to them.

CONFIDENCE

If learners are confident they can achieve the objectives set out by the learning, they are going to be motivated to do it. When previewing the objectives for the lesson, ask learners if they feel they can accomplish the work, then remind them at the end of what they have achieved. Create learning exercises or practice that inspire confidence. Don't make them too simple, but don't make them too hard either, except to challenge them. Try not to include too many exercises if you have a limited amount of time because learners may feel frustrated if they can't finish. Make note of individuals who seem reluctant or unsure and encourage and reward them for trying.

SATISFACTION

What are you going to do to ensure the ultimate satisfaction of the user with the learning? Candy may get their attention, but it won't satisfy them. If you make things relevant and build their confidence in the knowledge and skills, they will be satisfied. Satisfaction comes from having accomplished something worthwhile, so you should show them how the learning is worthwhile. Demonstrate what they will be able to do in the future with the new knowledge and skills. Remind them of where they started and where they have ended and ask if they are satisfied that they have learned something and can apply it.

In addition to those aspects, you may need to consider other aspects of learners' attitudes. In the instance of staff, their attendance may affect their pay status (promotion, raises, etc.). They may want some proof of their accomplishment, such as a certificate. Likewise for others, if they have paid for the sessions they may be very motivated to participate, but also want something tangible for their costs, such as proof of their accomplishment (e.g., introduction to a new search engine that has been proven to save time).

An example of accounting for learner attitude: Students required to sit in on a library information literacy lecture ("how to use the library catalog") often resent being forced to learn something which they feel is both boring and useless. Their attitude is quite evident in their faces and slumped body language. One group of instructors showed students job notices in their career fields and pointed to skills that centered around information retrieval and management. A link was made to demonstrate that the skills being taught were not only useful for a particular class, but for job placement enhancement and lifelong learning. The demonstration got their attention and established relevance. Throughout the session, examples related to job hunting were used, which helped ensure ultimate satisfaction for the students.

Why accounting for learner attitude is important: Some people argue that attitude shouldn't get in the way of learning, but it is a *factor* of learning. Perhaps an unwanted or unwilling factor, but it is a part of the whole package nonetheless. The better a trainer or teacher can understand and address attitude, the greater the chances of achieving successful outcomes for the learners.

TYPES OF LEARNING

In addition to analyzing learners, it is important to analyze the learning that is to take place. When it is understood what kind of learning will be undertaken, it is easier to match teaching strategies to learning objectives and learner styles, as seen in the chapter on developing learning. Here we take a look at applying analysis techniques to types of learning.

As set out by Robert Gagne in his book *Principles of Instructional Design*, it is convenient and useful to divide learning into the following five areas (Gagne and Briggs, 1974). By doing so, we can develop different teaching approaches for the different types of learning.

- Acquisition of motor (behavioral) skills
- Acquisition of verbal information
- Development of intellectual (procedural/conceptual/rule-based) skills
- Development of cognitive (analytical/problem solving) strategies
- Change of attitude (affective)

MOTOR (BEHAVIORAL) SKILLS

Motor skills are behaviors that can be physically taught, learned, or demonstrated—like moving a mouse, "right-clicking," or minimizing windows on the desktop. Motor skills are often assumed and taken for granted. Surely anyone can move a mouse, right? Older learners who haven't grown up with this ubiquitous input device for computers might not be able to do so with ease. Their physical movements, including focusing the eyes, may not be what they used to be.

Analyze learning situations to determine what motor skills are involved. For many applications in computer technology "mousing" skills are critical. But what about physically turning a computer around to troubleshoot whether the cables are plugged in?, or opening a printer and pulling out the toner cartridge? Look closely at how you teach someone these types of skills. For instance, one instructor has computer "newbies" pick up a mouse to look at and feel the ball that moves underneath. This gives practitioners a tactile feel for the mechanics involved in moving it around. Another instructor has learners practice highlighting text on a whiteboard to emphasize positioning the mouse cursor.

VERBAL INFORMATION

Verbal information often refers to facts and figures that are memorized. This includes definitions, terms, locations, concepts, meanings, directions, relationships, etc. Success in acquiring verbal information is measured when that information can be articulated and applied through recall, matching, sentence completion, etc. This is knowledge that is difficult to "do" in a hands-on lab. There are certainly many ways to facilitate the learning of verbal information—rote repeating, flash cards, question and answer, contests, etc.—that do not involve problem solving or intellectualization. However, both of those often rely on verbal information to complement or supplement learning. For instance, to use a search engine one has to remember the URL or where to find a link or bookmark to it.

Analyze the verbal information involved in a particular learn-

Fig. 1.11: Learning Types and Common Problems

Learning types	Common difficulties encountered
Motor Skills	• General hand-eye coordination • Maneuvering mouse precisely • Clicking or right-clicking fast enough
Verbal Information	• Memorization in general • Reluctance to use reinforcement techniques (e.g., flashcards) • Confusion with terminology
Intellectual Skills	• Remembering correct sequence of steps or events • Replacing old procedures with new ones • Disagreement with or lack of understanding of rules used
Cognitive Strategies	• Interpreting new situations and events • Applying two related procedures to solve a task • Unwillingness to try new ways (trial and error)
Attitude Skills	• Attitudes may affect others in class • May not be changed with training (e.g., related work schedule, pay, rewards, etc.)

ing situation (see Figure 1.11 for examples). If there is jargon, perhaps a list of definitions would be useful. Or maybe an illustrated guide is needed for the elements involved in an application indicating name and position. Potentially, anything you say in a lecture might be verbal information for which there should be a learning objective.

INTELLECTUAL (PROCEDURAL/CONCEPTUAL/ RULE-BASED) SKILLS

Intellectual skills include anything other than physical or verbal skills (or attitude, below). Anything to do with basic or low-level thinking is considered intellectual. This includes processes with a varying number of steps, from simple to complex. Learning to formulate a search query, filling out a book request form, and creating folders all involve intellectual learning. It also includes applying rules, such as for finding a book on a shelf by Dewey Decimal number, or applying concepts, such as identifying synonyms for a word or phrase. Much of traditional training falls into this category. Obviously, it will likely be matched with motor or verbal skills in any given application, but it focuses on the process of learning a skill or knowledge that involves thinking.

When analyzing an intellectual skill or set of knowledge, remember that all processes consist of steps. Usually there is either a literal sequence of steps involved, or there is something that has to be done first in a successive series of moves. For instance, finding a book can be broken down into a process (steps for searching the online catalog), concepts (organization on shelves by a location number), and application of rules (the Dewey decimal sequence). A literal step-by-step handout could detail the steps in searching, a graphic might depict the concept of number classification, and a guide to reading Dewey numbering could show how to interpret the numbering system. Each of these elements must be identified and tested—analysis should be conducted prior to developing a module.

PROBLEM SOLVING (COGNITIVE/ANALYTICAL) STRATEGIES

Cognitive strategies imply a higher-order level of thinking than simple intellectual endeavor. Rather than just "doing," cognitive skill implies figuring something out. Ultimately, cognitive skills and knowledge help anticipate scenarios, problem-solve, and develop alternative strategies to accomplish an outcome. The difference between cognitive and intellectual strategies is mainly creativity. For instance, an intellectual skill is the ability to highlight and "copy and paste" text in a word processor. A cognitive skill would be the creativity to try copying and pasting a file, based on an assumption that an entire file is simply a larger amount of digital information than a few words of highlighted text.

When analyzing for cognitive learning, don't be swayed to simply identifying solutions to examples. Most cognitive learning incorporates the same elements as intellectual learning. The difference is rather than achieve simple outcomes, cognitive involves more complex processes. Or, to use an analogy, an intellectual skill may be computing a sequence of numbers, but cognitive skill might be needed to determine where the numbers come from and verifying that they are correct. Intellectual skill implies a direct cause and effect ("when you get a 404 error message, click on the back button and select a different link"), but cognitive skill implies trial and error ("when you get a 404 error message, first try to reload the page a couple of times, then try editing the URL to get a higher level page in the directory"). For cognitive learning, identify the tools needed and strategies for using them.

ATTITUDE (AFFECTIVE) SKILLS

Although it isn't correct to label this "emotional" learning, it does pertain to things that affect the mood and motivation of learning. It relates not only to emotions and attitudes, but also to changes in preference or choice because of learning. For instance, by learning about certain customs, learners can appreciate why some people do things differently than others. Often as a result the learner seeks new applications or goals because of affective learning. And sometimes the affective learning is an indirect outcome. For instance, when Internet users understand what causes 404 error messages they are likely to become less frustrated when these are encountered, and thus the learners gain comfort as well as competence in using networked computers.

An example of determining different types of learning: When analyzing the e-mail training scenario from above, it is useful to divide the learning into its various types. Each type will have different objectives, different exercises and examples, etc. For instance, defining the concept of folders is verbal. Creating them is intellectual. Determining when to create new ones, or where to put them, would be cognitive. Actually "dragging and dropping" messages into folders involves motor skills. And when learners achieve time and energy savings, they may be motivated to start using folders in other areas, which involves attitude.

Why determining different types of learning is important: Not only do different types of learning require different methods of teaching, they require different approaches by the learner to acquire the knowledge or skills. By separating them into different categories you can help facilitate their learning. Often learners assume things are verbal that are really process oriented—it doesn't help to simply memorize the process, learners must understand the concepts and rules that apply to the steps. Sometimes learners (and teachers) assume motor skills outcomes are intellectual-based and get frustrated when they can't "figure out" something that is a matter of hand-eye coordination. For instance, finding a menu option to carry out a task is intellectual, but involves prerequisite motor skills (such as clicking on an item to select it so that the option is available to use on the object). Separating and categorizing the learning is truly a matter of facilitating the learners who will eventually learn how to learn better.

STYLES OF LEARNING

There are many ways to assess learning types, ranging from quick-and-easy, rule-of-thumb approaches to in-depth assessment tests. These latter are likely to be most helpful in diagnoses of young learners who will be in a classroom with the same teacher for an extended period of time (especially during the student's formative development). However, for many teachers and trainers of adult learners, it is helpful to have a cursory understanding of the styles of learning. This is something that is difficult to assess for each learner in a given situation. It is best to be equipped by knowing what learning styles you might encounter and then developing strategies to deal with them. As this involves data gathering and looking at trends, it is a form of analysis. When you encounter a learner who has a particular preference you will be able to address it.

Three areas of learning styles to look at are:

- Auditory, Visual, and Tactile
- Deductive and Inductive
- Abstract, Concrete, Reflective, and Active

AUDITORY, VISUAL, AND TACTILE

Quite often, learner styles are categorized as auditory, visual, and tactile. This is oversimplified many times as people who learn by hearing, seeing, or doing. An auditory-learning-oriented person is someone who not only learns by listening, but also actually gets validation externally from other people. This includes the instructor, but other participants as well. A truly auditory person may sit in a classroom staring at the ceiling or at the floor, absorbing every word that is said. Or he may turn around to listen intently when others speak.

A visual-learning-oriented person is someone who not only learns by seeing, but also needs to see a graphic depiction to learn. Graphs, charts, and conceptual diagrams help reinforce learning for a visual person. Even text is often treated as an image. This learner will likely read a screen instead of listening to the instructor. Or she may jump ahead to "pictures" and ignore text altogether. She may focus on scanning the handouts rather than listening solely to the instructor.

A tactile-learning-oriented person needs to do, to practice something to reinforce learning. Hands-on is as much a credo as it is a preference. Practicing may take the form of working through steps or trying a process several times. The act of performing allows her to see and understand what's going on.

The analysis of these different types of learners doesn't lend itself easily to data gathering and looking for patterns and trends. Rather, it requires that teachers be aware and know what to expect in preparation for learning experiences (see Figure 1.12). There are strategies that lend themselves to addressing each of these preferences to facilitate learning. For instance, asking others to speak.

Fig. 1.12: A/V/T Learners and Ways To "Reach" Them

Learner	Approaches
Auditory	• Audio stimulus (speaking, recordings, reading aloud) • Multiple voices (participants speak up, answer, discuss) • Verbal clues ("Pay attention to this next definition.")
Visual	• Visual stimulus (text, images, animations) • Express concepts in terms of concrete objects (e.g., e-mail messages are like letters inside of envelopes with addresses) • Visual clues (arrows, blinking/underlined text, circled objects)
Tactile	• Kinesthetic stimulus (handling, seeing up close, doing) • Hands-on practice • Tactile clues ("Pick up the mouse and feel the underneath.")

Fig. 1.13: Deductive/Inductive Learners and Ways To "Reach" Them

Learner	Approaches
Deductive	• Elaborate on concept or rule • Provide challenging exercises to apply rules
Inductive	• Mention concepts while describing procedure • Provide routine exercises to practice procedure • Ask learners to describe concept/rules in their own words

DEDUCTIVE AND INDUCTIVE

Probably lesser known than the auditory, visual, and tactile preferences for input are two types of logic styles that also affect learning (see Figure 1.13). Deductive refers to making deductions, that is, taking a general principle and determining examples and applications from them. For instance, given the principle that "anything can be cut and pasted," deductive learners would accept the generalization and assume that they can use it whenever needed. They might not need to practice it, or if they did, they would like trying a wide range of possibilities.

Inductive learners, on the other hand, work from the opposite direction. They can take a handful of examples and figure out the principle. They may experiment or accidentally try several variations that lead them to understand what's going on. Having cut and pasted text, they may realize that the cut and paste option seems to be available in several other situations. Only after they have tried it will they come to the conclusion, "anything can be cut and pasted."

Fig. 1.14: Matrix of Perceiver vs. Processor Learners

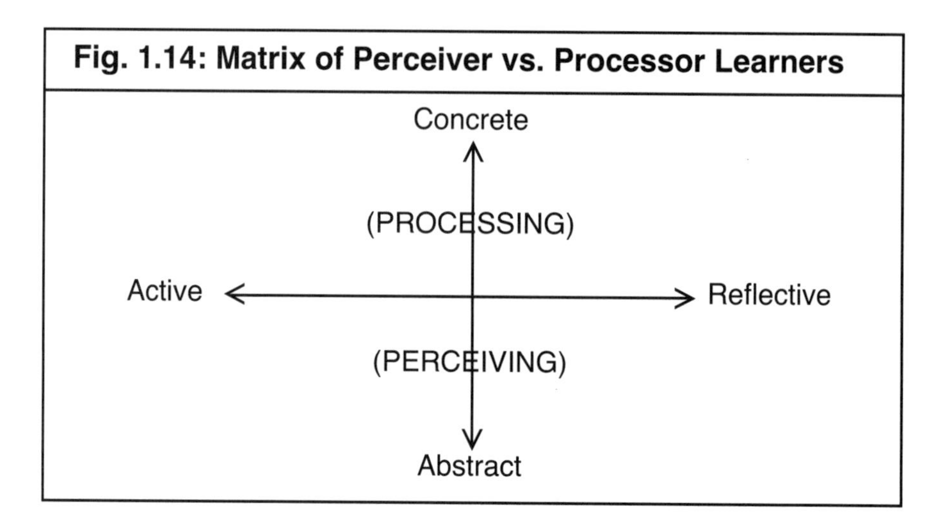

ABSTRACT, CONCRETE, REFLECTIVE, AND ACTIVE

Another way to classify learners is to make a distinction between how learners perceive and how they process learning (see Figure 1.14). Originally, David Kolb set up a distinction between abstract and concrete perceivers and reflective and active processors in his work, *Experiential Learning* (Kolb, 1984). Though similar to classifications above, this view is presented to give a more complete picture, and further insight into avenues of analyzing learners.

Perception preferences relate to the learners' tendencies to prefer different types of learning input (see Figure 1.15). Abstract perceivers tend toward intellectual stimulus. They relate to analogies and analytical presentations that may let them think about what's going on. Concrete perceivers prefer things in ways that deal with the senses and involve doing. They relate to hands-on activities, role-playing, and discussions.

Processing preferences relate to the learners' tendencies to use different types of processing of learning. Reflective learners tend to internalize, step back, and think about learning. They often relate to lectures, reading and writing summaries, or reports. Active processors react to learning. They relate to interactive sessions and immediately practicing what they've learned.

An example of accounting for learner styles: As noted, it is often difficult to assess the learners prior to a learning situation. But by understanding preferences, you will strengthen your ability to help learners. When you see someone go blank as you explain an abstract concept, such as how a "pop e-mail client" works, you might guess that they are not an abstract perceiver. Your lesson plan should include compensation for the other type

Fig. 1.15: A/C/R/A Learners and Ways To "Reach" Them

Learner	Approaches
Active /Concrete Perceivers	• Allow doing, applying • Independence, Self-Discovery
Concrete/ Reflective Perceivers	• Allow analyzing, integrating experiences • Discussion, Debriefing
Reflective/Abstract Processors	• Information gathering, research • Instruction, Reading
Abstract/Active Processors	• Problem solving, practice applying concepts • Coaching, Guiding

of perceivers, perhaps by having a diagram of how e-mail gets from the server to the individual's computer. Likewise, try to build into the learning compensation for other differences. Use visuals, but encourage discussion. Identify principles, but allow hands-on time to work through examples. Realize that some people may prefer to make notes than do the hands-on practice, or they might want to watch and ask questions while others do it. Try to account for variations in learners in the learning situations.

Why accounting for learner styles is important: Whether looking at types of input (auditory, visual, and tactile), problem solving (deductive and inductive), or perceiving/processing (abstract and concrete, or reflective and active), learners have a variety of preferences. As noted, some of them seem opposite of each other. Be careful not to plan for only one type of learner—we often tend to create learning situations that favor our own styles! Try to include variety to include as many different types of learners as possible.

REVIEW TEMPLATE FOR ANALYZING LEARNERS

Scan the following as checklist for comprehension of this chapter on analysis. Also, you can use this list as a quick checklist when you are working through analysis for your own technology teaching.

How will you determine learners' needs?
- What are their basic, or general, needs?
- Where are their current skills, and where should they be? What's the gap?
- Can you identify a possible outcome to gauge or learning hypothesis to test?
- What method will you use to make inquiries?
- What questions might you ask? Whom will you ask?
- How will you collect the data or information? How will it be organized?
- How can you get user input into the analysis process?
- What will you look for when you analyze the data/information? How will you summarize what you've found?

What level of experience do the learners possess?
- Which knowledge or skills are affected by level of learning?
- What kind of questions can you ask to identify levels?
- How can you word questions so responders are comfortable in their responses?
- When and where will you ask?

To what extent can you identify learner attitudes?
- Are there any underlying motives or reactions to the learning?
- Are learners there because they want to be, or is it mandatory?
- How will you get the learners' attention? How will you make the learning relevant for them? What will you do to help build their confidence in the learning? What can you do to ensure satisfaction?

Where will you include different types of learning?
- Are there any motor skills involved in this learning? If so, to what degree have learners already mastered them?
- Can you identify verbal information that must be acquired?
- What intellectual skills are involved? Can you identify the steps required to work through them?
- What cognitive skills are involved? Can you identify prior intellectual skills needed to accomplish the learning? Can you identify problems for hands-on practice?
- Are there learning attitudes involved? What are they and how do they apply? What outcomes will be effected by them?

How will you consider different styles of learning?
- Where do auditory, visual, or tactile input preferences fit in?
- How is deductive versus inductive learning involved?
- Where are differences in perceiving or processing noticed?

Can you accommodate for all of these?
What are your preferences? Which preferences are you most likely to overlook?

Which category of analysis will you use?
What forms or surveys will you use for a direct-formal analysis?
Who will you contact if performing an indirect-formal analysis?
When and where will you do informal analysis?
How many people will you survey or talk to?
Is there a reason to combine types of analysis?
What tools will you use to gather information?

REFERENCES

Gagne, R. M., and L. J. Briggs 1974. *Principles of Instructional Design* (2nd edition). Fort Worth, Texas: Holt, Rinehart and Winston.

Keller, J. M. 1983. *Development and Use of the ARCS Model of Motivation Design* (Report No. IR 014039). Enschede, Netherlands: Twente University of Technology. (ERIC Doc. Reproduction Service No. ED 313001).

Kolb, D. A. 1984. *Experiential Learning: Experience as the Source of Learning and Development*. Upper Saddle River, N.J.: Prentice Hall.

2 STEP 2: DESIGN

In this chapter you will:

- **Describe five aspects of designing learning**
- **Define what learning objectives are**
- **Discern four elements of an objective**
- **Describe how to build objectives**
- **Determine prerequisite learning for an objective**
- **Determine learning outcomes for an objective**
- **Identify nine uses for objectives**
- **Review template for designing objectives**

INTRODUCTION

There's an old adage about how to give a successful speech, "Tell them what you're going to tell them; tell them; then tell them what you've told them." It presumes that you should give your audience a context or idea of what to expect, deliver your ideas, and then reinforce the message by highlighting it again. It should be emphasized that this is an old adage, and that it is only useful for giving a speech. Unfortunately, some people think it applies to teaching as well.

What is teaching, after all? Sharing knowledge, right? And done verbally in a classroom, it is a lecture or speech. But it has been proven that a lecture is one of the least effective ways to teach. People don't learn much from a speech or lecture. Usually they write down notes, and in the process of actively deciphering information they may be stimulated to learn, or maybe only to memorize. Reviewing the notes later helps to reinforce and retain the learning.

So if giving a lecture isn't the best way to teach, what is? Usually it takes a combination of elements to facilitate effective learning, which includes: gathering appropriate content, employing well-structured development, and utilizing effective methods of delivery. The way these elements are put together shall be referred to here as designing learning.

What's the difference between designing and developing learning? Designing is an intellectual phase often likened to an architect creating blueprints that are later used by the construction company to literally build a house or building. Design is making abstract decisions (why) and development is making concrete decisions (what, where, which). The two are very closely related. Sometimes you might move back and forth between designing and developing, and at other times you might do both simultaneously.

DESIGNING LEARNING

The basic method for designing learning put forth here has four main aspects:

- Describe expectations explicitly and concretely
- Outline previous applicable learning
- List precisely what should be learned (objectives)
- State outcomes that are measurable

In a sense, you could say: show the learners what they are going to do, help them do it, and have them prove they have done it. As will be shown later, it is helpful to start with objectives first, and then complete the other parts afterwards. In addition, there is a fifth aspect that is very closely related to design but is really a part of development:

- Determine methods to help facilitate that learning

EXPECTATIONS

It is important to identify expectations for learners (see Figure 2.1). They need to know what is expected of them for a specific learning situation. If they need to prepare ahead of time, they should be advised. If they will be performing hands-on exercises and their work will be reviewed, they need to be told. If they are required to complete some kind of homework after the class, they should be warned.

An example of using expectations: When posting a description of a class or course, include information regarding what you expect learners to do, or not do. They may think that they can sit there and listen to a lecture. But you might want them to do some hands-on searching. You might let them know that it is okay to work in pairs. Or you might demand that they not check e-mail during the session. Note: it is better to let them know sometime before class, rather than at the very beginning of the session.

Why describing expectations is important: People need to feel comfortable when they learn. It helps to give them a context.

Fig. 2.1: Sample Learner Expectations

Course	Expectations
Introductory E-mail	This beginning e-mail discusses general policies regarding e-mail and teaches basic skills in reading, replying, composing, and sending e-mail. It will cover how to organize an inbox by deleting mail, creating folders, and moving messages. Participants are expected to engage in all activities using a separate e-mail server (i.e., you will not be able to read your own mail in this class).
Internet Searching 2	This Intermediate Internet Searching class will compare searching in five different search engines (Yahoo!, About, Alta Vista, Google, and Fast). Participants will be expected to work in pairs to formulate and revise search queries, perform searches in all search engines, and discuss results and findings with the class.
Troubleshooting	This course is geared for new employees to provide them with skills to do "first level" troubleshooting hardware and software. Participants will be expected to practice: unplugging and plugging cables into the back of a computer; copying error messages; using Task Manager to end applications; restarting the computer.

Without the expectations learners may be confused about why they are there, and not find enough motivation to make the learning significant.

PREVIOUS LEARNING

Previous learning that was required or has been covered should be outlined (see Figure 2.2). Learners need to know where the starting point, middle, and ending is in the learning. The starting point includes both the literal beginning of the learning and some idea of what came just before the beginning. This is often referred to as entry-level behaviors/skills or the prerequisite knowledge needed to undertake a particular learning. This not only gives the learners a context, but also helps them self-assess to determine if they are ready for this learning circumstance. As will be shown under objectives, the expectations can be expressed as previous lessons or earlier verbal knowledge.

An example of outlining previous learning: One example in particular comes up a lot in workshops. When a class requires hands-on participation, a prerequisite is that learners can use a mouse efficiently. State clearly that participants in a class must be able to use a mouse proficiently because time cannot be taken during class to help them. (Presumably there is a mouse class to

Fig. 2.2: Samples of Previous Learning

Course	Prerequisite
Introductory E-mail	Prior to this class participants must: • Be able to log in to access the network • Have an e-mail address • Opened Outlook and read pgs. 2.1-2.14 of Outlook workbook
Internet Searching 2	Prior to this class participants must: • Have taken (or "tested out" of) Internet Searching 1 • Be Web browser proficient (or taken Intro to Netscape) • Review "Anatomy of a Search Engine Index" handout
Troubleshooting	Prior to this class participants must: • Be able to log in to access their computer • Have experience working with Windows and at least two applications: Word and Netscape • Review list of "Computer Terminology" handout

which you can refer them!) As with expectations, it is most useful for the learners to know this before they take the class.

Why outlining previous learning is important: Learners need to understand why they are starting at a certain point. They need to be able to determine if they are ready for the learning. Their motivation and attention become stronger if their confidence is built up.

OBJECTIVES

In describing exactly what the learning comprises, it is very useful to utilize a structured learning objective. This tool allows you to state exactly what the learning covers. You can state to whom it is aimed, on which actions or behaviors it focuses, any conditions on the learning, and to what extent or degree it extends. This is further detailed under the section on objectives below.

An example of listing objectives: In the author's experience, it is best to simply provide a list of objectives as a handout of the class. These could be simple: "Learners will be able to choose a search engine based on their topic." Or they could be complete: "Freshman students will be able to brainstorm keywords for their topic and use a list describing the subject areas of several search engines to create a match that produces two results that meet search criteria."

Why listing the objectives is important: The knowledge you are helping them to learn is not supposed to be a secret. Be upfront. Be obvious. Share. Tell them what you are going to tell

Fig. 2.3: Samples of Outcomes

Goal	Measurable Outcome
E-mail management	Less in-box messages; More folders; Folders grouped by topic
Internet searching	Description of differences between search engines; Successful searches; Referrals to appropriate resources
Troubleshooting	Less downtime; Less calls to Help Desk; Less frustration

them. As noted below, there are several uses for a list of objectives, for them, and for you (as well as others).

OUTCOMES

The learning objective must somehow be applied, and there should be demonstration or proof that learning took place. This is sometimes referred to as the outcome (see Figure 2.3). It is used specifically to mean evidence, or proof, that learner has learned. This should be demonstrated in some way, an application of the skill or knowledge.

An example of stating outcomes: Simply, you should be able to tell the learner how the learning will be demonstrated. For instance, for an Internet searching class you might state: "Within a week after the class, learners will be expected to fill a sheet (or online form) detailing the following: a search topic and keywords associated with it, description of which search engines were used (and why), outline of the type of things found in the ensuing results list from the search and a revised query (as needed), and a description of a result that met criteria of the search."

Why stating outcomes is important: How do you know if someone has learned? If you ask them, and they nod their heads, does that mean learning has taken place? Maybe, but probably not! It could mean they are tired, or embarrassed, or understand in an incomplete way. Nodding doesn't demonstrate the learning.

METHODS

Once it is decided what will be learned, what needs to be known prior, and how the learning outcome will be demonstrated, the next step is to decide how the learning will be facilitated. Primarily this means using teaching techniques or strategies to deliver the lessons. Various methods or approaches can be used, but the ones chosen must best reinforce the learning. For example, analogies may be effective ways to explain concepts, but a hands-on exercise is probably not a useful way to learn vocabulary. For our purposes, this is technically defined as part of the develop-

Fig. 2.4: Use of an Analogy

	Postal Mail	E-mail
Recipient	Includes person's name, street address, city, state, and zip code	Includes e-mail name, @, and domain address for mail
Sender	Includes similar info as recipient	Automatically attached
Body	Written on paper, enclosed, and sealed in an envelope	A text or HTML file that follows the address and header
Attachment	Usually secondary letters, etc.	Any file you want to send
Sending	Drop into a post office receptacle for pickup, processing, and routing	Sent though mail server and routed through Internet to destination

ment of learning, and is left to the next chapter. However, it is mentioned here to emphasize the interdependent relationship of designing and development.

An example of determining methods: Using analogies is one method for explaining something which is abstract, such as the Internet (see Figure 2.4). An analogy like "the information superhighway" works only when it relates directly to the subject at hand. It might be useful when discussing routers and relays, but not e-mail. An analogy like "an information mall" might be useful when discussing the variety of search engines available and the variation in the quantity and quality of their contents.

Why determining methods is important: Methods are the tools used to deliver the content for learning. As noted earlier, a verbal speech isn't very effective. Other methods such as hands-on demonstration, "show and tell," active participation and group learning are far more effective in helping learners learn. But each method should be used in the right context—group learning may not be useful if participants need to have their own computer for practice.

The most common failure in designing learning is the lack of a specific plan of attack that combines all of these aspects together.

DEFINITION: LEARNING OBJECTIVES

Learning objectives describe a measurable accomplishment that is outcome-driven. That is, the learning is supposed to utilize a process that yields an easily recognizable result. As such, a learning objective is a tool for organizing units of learning. A lesson plan will likely have several learning objectives. Each objective describes the content and scope of a particular piece of the learning. Objectives are supplemented by expectations, supported by previous learning, delivered using a method or strategy, and measured by their resulting outcome.

Each objective has several components. It may be useful to identify who will be engaging in the learning (participants). It is crucial to explain exactly what skill is to be learned with each objective (behavior). Because a given skill could be performed under various circumstances, it is necessary to describe the conditions associated with the objective (circumstance). And because the skill can be formed to a certain extent—from perfunctory to mastery—it is important to list the degree to which an objective should be performed (degree).

It is possible to teach without using learning objectives, but teaching without such a structure leaves a lot up to chance. Objectives spell out all the components in a single, condensed statement. But more than that, the system helps build learning by using principles that result in instructions that are clear, precise and measurable. As will be suggested below, it is often difficult and not always easy to include all of these components. Learning that does not include them can suffer by being incomplete, vague, or confusing. That doesn't mean that learning will fail if they are not used—it means you may be leaving something open to interpretation (or chance). If you learn the system outlined below, and use it whenever you can, you will grow to appreciate and rely on it.

PARTICIPANTS

Most designers aim their instruction at the group of people with whom they directly interact. Schoolteachers design for students at the grade level they teach. College instructors do likewise. Lifelong learning associations design for senior citizens. Industry designs for specific employees. Why should we explicitly state for whom the learning is intended? Two reasons: First, within a group it is very likely that there are individuals of varying levels of experience, skill, or expertise. Second, the instruction might be transferred from group to group and should indicate for what level it is designed.

Thus, participants can be defined by their nature or composition, and by their level of experience or knowledge. The composition defines who they are based on category: occupation or vocation, rank, age group, etc. For instance, in an academic setting there is a difference between a "true freshman" (first year after high school), a program freshman (someone who transfers to a new program but may be in her or his third year of college), and a returning student freshman (older or "adult learner"). Each of those categories likely relates to varying levels of study experience and motivation. Similarly, differences in age groups may include varying life experiences, in addition to overall education. Those differences can make a big difference in expectations, outcomes, or strategies used to facilitate learning.

Explicitly stating level of skill, knowledge, or experience needed is very important. Identification of prerequisite skills, discussed below, helps to identify what the learner should know prior to the beginning of the new learning. Objectives should state that these prerequisites are mandatory. (Even so, quite often learners of varying degrees of experience are lumped together. Approaches for dealing with these are covered under the development of strategies to facilitate learning.)

Sometimes objectives do not state for whom the learning is intended because the scope of participation is narrowed by confines of participation. For instance, it seems redundant to state to a group of first semester college freshmen that take communication courses in their second semester that a learning objective is for "first semester college freshmen, regardless of skill level, prior to taking COM101." However, it is better to be explicit, even if it means being redundant, than to be vague.

An example of identifying participants: Many public librarians still describe the need to work with older learners who are not proficient at using a mouse to maneuver around the screen. It is important to explicitly state what proficiency at "mousing" means—what skills and knowledge (understanding of terminology, elements of the screen, etc.) are required. For instance, mousing can be described as "the ability to move the cursor around on the screen, including double clicking and right clicking, to select icons, open and close windows, highlight to cut and paste, and enter text into text boxes." Thus, you would describe participants as "having mouse proficiency" and define the term for them.

Why identifying participants is important: Among other things, when instruction has to be constantly interrupted to assist learners who do not have prerequisite skills, the entire class suffers. Or, if it is online, unprepared learners soon become confused and lost. Likewise, learners who exceed the minimum entry skills may soon become bored and frustrated. By stating exactly for whom the course is designed, learners are better equipped to assess themselves and their participation in the learning.

SPECIFIC BEHAVIORS (SKILLS)

The crux of the objective is to state exactly what skill, behavior,

Fig. 2.5: Example of Measurable Action Verbs

Example of measurable action verbs				
Affective	**Motor skills**	**Verbal knowledge**	**Intellectual skills**	**Problem solving**
Enjoy	Drag	Define	Select	Apply
Encourage	Click	Explain	Identify	Determine
Feel (comfort)	Move	Describe	Complete	Hypothesize

or knowledge is required to accomplish the learning objective. As noted previously, it is helpful to first define whether the learning will fall into the domain of motor skills, verbal knowledge, intellectual skills, or problem solving. Once the learning domain is defined, the specific actions can be stated. A learning objective statement could start with the phrase "learner will be able to," followed by a measurable action verb which comprises the activity, and accompanied by conditions and degrees as outlined below.

Stating the objective with a measurable action verb is critical. Learners must be able to demonstrate the skills listed. If it can't be demonstrated, it can't be measured. If it can't be measured, we can't know whether learning has been accomplished. Some examples are given in Figure 2.5. The measurability, if you will, of the objective leads to instant recognition by the learner that she or he has learned something. And it can be easily and directly converted into test item for evaluation of learning. Take, for instance, the objective: "When viewing a Web page longer than 28 lines, a learner will be able to demonstrate four methods for scrolling down the page." Learners can immediately show they can perform the tasks, once shown the methods. And a test item is easily generated: "Describe four ways to scroll down a page."

Many people think that a learning objective is a concept, not a tool, and believe it's okay to use non-measurable, non-action verbs. Don't do this! Take for instance, the objective "Learners will understand they can scroll down a page several ways." Understand is not an action verb, and in itself is not measurable! How do you measure understand? When you ask someone, "Do you understand?" and she nods her head yes, you have no demonstration other than nodding. The objective is not supposed to be, "When asked whether she understands the lesson, the learner will nod her head." *That* is all that's been accomplished. How is understanding demonstrated? How is it tested? By using measurable action verbs, the learners (and the teacher) know exactly where they stand.

An example of stating specific behaviors: The act of scrolling down a screen may seem trivial, but it's not, really. Tests show that a lot of people don't look past the first screen of a Web page. There are a number of reasons for this, but if the learner has a full arsenal of options (including shortcuts) it is more likely that she or he will use them, and gain the benefit of using them. Some people don't like to scroll because they simply click on the scroll bar arrow and it takes too long to get anywhere. Are they aware that they can click on the scroll bar and move up or down a page (or chunk of a page) at a time? Or that they can grab the bar on the scroll bar and drag it quickly to the end of the page? Or that they can use the Page Up/Page Down keys and move quickly without using the mouse? For all of these reasons, it's a good idea to teach new users how to scroll.

And the objective for this could be: "When faced with a page that extends beyond one screen, new users of computers will be able to scroll up and down a page by: 1) clicking on the scroll bar arrows to move small distances; 2) clicking on the scroll bar to move in larger chunks (above the bar to move up, below the bar to move down); 3) grabbing the bar on the scroll bar and pulling/pushing to move quickly up and down a page; 4) using the Page Up or Page Down keys to move in large chunks through a page."

Why stating specific behaviors is important: There are three purposes behind using measurable action verbs to represent and accomplish an objective. First, it forces the teacher to express learning objectives in simple and concise language that is not vague or easily misunderstood. Structure is a good thing! As will be shown, breaking things down in structured units helps organize and maintain learning. Second, it gives the learner something very concrete and direct to learn. There should be no misunderstanding on the learners' part about what is being learned. Third, the objectives lead directly to an evaluation item covering that learning. Realistically, it's not always possible to test learners in all situations. But the test should be there in case the teacher needs to justify that learning has been done, or the learner needs the feedback and reinforcement.

CIRCUMSTANCES

Also known as the conditions of learning, or the environmental influences under which learning takes place, the circumstances describe the context of why, when, or where the learning happens. What causes this learning to be needed in this situation? As noted in the example above, the need for scrolling occurs when a

Fig. 2.6: Example of Degree of a Circumstance

Behavior	Possible degrees
Search	• By using operators (+, –, “”) to compose query • And formulate an alternate query using narrower terms
Revise search	• Finding additional (or fewer) results • Until a specific resource is found
Evaluate	• Find 5 elements to establish objective relevance (authority, bias, date, source, reliability) • Find 2 elements to establish subjective relevance (complements topic, appropriate length, etc.)

page extends beyond one screen. Without this condition, the objective lacks a context and is in danger of being too abstract.

An example of establishing circumstances: For every learning objective there should be an explicitly stated circumstance describing why, when, or where the learning happens. If the learner is being told to type in a URL to get to a search engine, the circumstance might be, “when retrieving a search engine home page, learner will be able to type…”. Or it might be “where given a list of URLs to choose from,” or it could be, “given the name of a search engine, ‘guess’ the search engine’s URL by adding www. before it and .com to the end of it.”

Why establishing conditions is important: Learners need context to make learning easy to understand. Objectives need context to state specifically the conditions under which this learning happens. And the context may change in different circumstances. Typing or editing a URL may be set in the context of spelling things correctly to make novices feel comfortable with typing, or to teach problem solving when a link doesn’t work (e.g., returns an error) to intermediate learners. Circumstances set up a conditional prompt in learning where learners can ask themselves, “When do I apply this learning?”

DEGREE

Okay, so you have a behavior and the conditions under which it will be undertaken. But to make it easily measurable, limits need to be established for the action. The condition indicates how much, how long, how far, etc. (see Figure 2.6). The degree gives the objective added measurability, contributes to the context, and makes it “real.” For instance, when learners modify a URL, to what degree should it be modified? Should it be completely deleted and retyped? Should it be edited to result in another Web page? The degree does not indicate how to do the behavior, but the extent to which it will be done.

An example of setting degree: Every learning objective should have a degree. Whatever the behavior, it has to be done to some resulting degree. In the example from above, we want learners to scroll down a page in four ways, but we don't say to what extent. A condition might be, "To display parts of the page which are 'off' the screen." The lack of a condition can leave learners feeling that the objective doesn't have a point. To say that learners simply need to be able "to scroll" seems pointless.

In some situations, the degree is either built into the objective, or it is understood that the degree requires mastery of the objective. For instance, "scroll" pretty much implies displaying the rest of a page or document. "Open a second window" pretty much implies that the learner will be able to view other information in that window. Something, like "define the topic" implies that the behavior is carried out until the definition is complete. "Restart the computer" indicates that the computer will be restarted by the learner. Thus, if you initiate the action it will go on until completed. When the action requires the learner to continue until complete, the degree is said to be mastery level. Still, to be helpful and precise it is useful to include the conditions.

Why setting degree is important: Learners need to feel that the objective is purposeful. The perception that they are simply doing something for the sake of doing it can undermine motivation. But the degree also helps give parameters to the measurability of the objective. It helps define the extent (how far? when will I know I'm done?) to which the action is to be achieved. Some objectives even go as far as to say, "90 percent of the time." This degree gives the learner an idea of the level or extent that needs to be demonstrated.

When using a four-part objective, several of the parts of designing learning become clearer. Given the overall objective, the expectations for learning become evident. Having established precisely what is to be learned, the previous learning and outcomes are easier to determine. And possible methods for facilitating the objectives begin to make themselves apparent. Before we go on, we should practice building these objectives.

BUILDING OBJECTIVES

To design objectives you need to investigate several aspects involved in the complete process. Too many times the participants' perspective is overlooked, as well as the circumstances and degrees of the learning. The formula is a reminder to gather information about the learning that will help you put together effective learning objectives. After all, effective objectives lead to effective learning.

$$\textbf{objectives} = \frac{\textbf{participants + skills + circumstances + degree}}{\textbf{type}}$$

The four parts of the objective are important, but must be balanced over the type of objective being facilitated. Let's look at five different types of learning scenarios and walk through how we would build learning objectives:

AFFECTIVE

An affective objective involves a situation where you want to change an emotion-based behavior. Usually in technology training the affective area gets overlooked, and yet it is common to see people frustrated or confused when using computers. Many introductory courses spend time with novices, trying to make them feel more comfortable or confident using computers. You can't say simply, "Learners will learn to be more comfortable." However, you can identify behaviors to work on that cause frustration or confusion.

Take for instance, using the mouse. Handling one requires dexterity that is often at odds with older hands and eyes. (In fact, it seems that a mouse is made for the young—why isn't there an alternative for the elderly?) A learning objective for older computer users that has them using the mouse to select, highlight, and drag options has as its goal changing how they feel (comfort, confidence, ease). The circumstance for such an objective would likely relate to something that would normally cause frustration. The degree would likely indicate comfortable success, not precise achievement.

> "While typing a URL [condition], a staff member [audience] will move the mouse cursor slowly within the text area and click once to practice typing text into it [behavior], making as many mistakes needed until s/he feels comfortable with completing a URL [degree]."

MOTOR SKILLS

Motor skills are something that adults often take for granted because they are more often associated with the development of children. However, anytime physical dexterity is required as part of learning, motor skill objectives are needed. As in the case above, learning the motor skills associated with using a mouse is something that many new computer users may encounter. The participant, behavior, and circumstance may be the same as above, but the outcome is to achieve precision or mastery.

> "When presented with a textbox on a Web page [circumstance], a new computer user [participant] will move the mouse cursor within the text area and click once to begin typing text into it [behavior] quickly on the first try [degree]."

Note the difference between degrees. The phrase "with ease" is meant to allow the participant to take his or her time. It might even carry a further degree, such as "60 percent of the time." Part of the intent is to allow the participant to build up confidence. The phrase "quickly on the first try" implicitly sets a time limit (hurriedly) and a limit (once). The goal is to improve dexterity.

VERBAL KNOWLEDGE

Verbal knowledge is sometime called, "things you have to memorize," and often forms the majority of objectives in lecture-based learning. However, to state that "learners will know the answer to" doesn't use measurable action verbs. Identify the outcome or outward manifestation of memorization. Action words that could be used include: define, list, describe, enumerate, state, recite, and draw. The verb match could also be used if the learner is matching a term with its definition.

Verbal learning objectives are especially useful for learning that doesn't include hands-on involvement. If someone can't demonstrate where an option is on a menu, they should be able to match the option name with the appropriate menu item.

> "Given a list of options and menus [circumstance], first time users of Windows [participants] will be able to match [behavior] seven options with their corresponding menus successfully [degree]."

INTELLECTUAL SKILLS

Intellectual learning objectives are probably the largest category of objectives, especially in technology learning sessions with hands-on activities. Intellectual implies making decisions or choices and choosing options. It emphasizes that learners use intellect to guide them to an outcome. Whereas motor skills tend to be purely physical and verbal ones are verbal, intellectual skills combine reason (applying an option) with motor (selecting with the mouse) and verbal (knowing the definition of a menu option) skills.

> "Given the need to double-space the text in a document [circumstance], intermediate users of Word [participants] will be able to highlight text and select [behavior] the "Paragraph" option and change the "Line Spacing" attribute to "Double" to double-space the text [degree]."

PROBLEM SOLVING

Problem-solving objectives go a step beyond intellectual ones. The difference is that instead of simply choosing an option, learners must determine which option applies. For instance, rather than simply define a menu option, a learner would be asked to differentiate between two similar options and apply the more appropriate one. Literally, the objective looks at a problem to be solved rather than a straightforward task that needs to be completed.

> "Having revised a search [circumstance], novice users of *Yahoo!* [participants] will be able to determine [behavior] which links, if any, match their search criteria [degree]."

Notice that all of these objectives are precise, if not concise as well. Each of the four parts of the objective makes it explicitly clear what the learning covers. A starting point, or prerequisite learning, is hinted at, but should be made clear. Likewise, it should be fairly evident what the outcome should be, but again that needs to be spelled out.

DETERMINING PREREQUISITE LEARNING

Simply stated, each objective is likely to be the prerequisite for the next learning objective. In fact, once you have built up a system of objectives over time, this will likely be the case. And hope-

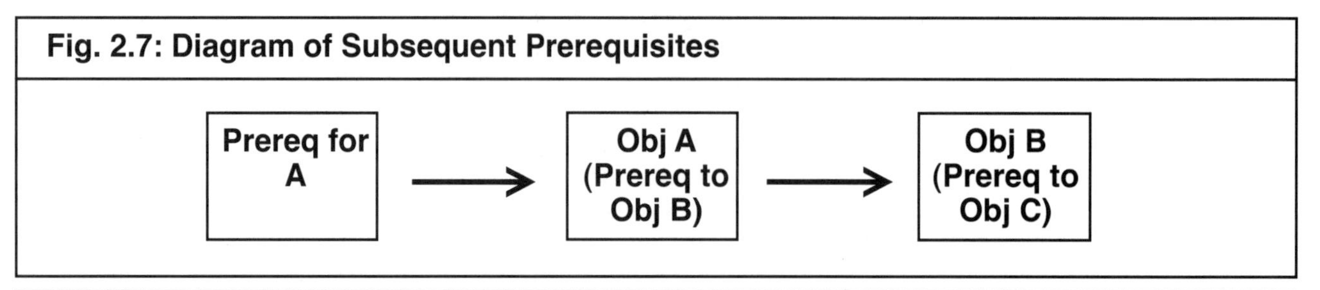

Fig. 2.7: Diagram of Subsequent Prerequisites

Fig. 2.8: Sample of Successive Prerequisites

Mousing Skills

- Move mouse cursor
- Point & click
- Click & drag
- Right click
- Highlight

Windowing Skills

- Open & close windows
- Scroll up & down
- Open multiple windows
- Move between windows
- Cut & paste

Web Browsing Skills

- Click on links
- Type & edit URLs
- Open multiple browsers
- Reload pages

fully you can make the prerequisite objectives from previous learning mandatory for the lesson at hand by literally requiring learners to complete the previous lessons (or somehow demonstrating the entry level behavior needed) (see Figure 2.7). However, making prerequisites mandatory is not always possible, although everybody needs to start somewhere.

For any objective that does not have a prerequisite lesson, the prerequisite learning must be identified and articulated. The best way to do this is look at the current objectives and identify the steps involved in completing the objective (see Figure 2.8). Given the steps that will comprise activities in the learning, ask yourself "What does someone need to know to do this?" Don't take anything for granted. List the verbal, motor, intellectual, and problem-solving skills needed.

An example of determining prerequisites: Let's look at an example of a basic course, "Introduction to Browsers." The goal of this course is gaining familiarity with key parts of a browser, such as basic menu options, property settings, and the bookmark manager. The objectives are going to deal with verbal and intellectual skills, but involve a lot of motor skills (pointing, clicking, right-clicking, selecting) that are assumed. However, we need to explicitly state what those skills are:

> "Prerequisites for this course include the ability to use a mouse to click, right-click, or use the menu bar to select menu options; open, drag, and move between multiple windows on the screen; and highlight, cut, paste, or drag text. If you do not feel comfortable with these prerequisites, please see the instructor about ways to meet these required skills."

Why determining prerequisites is important: First and foremost, prerequisites lay out what is required to undertake learning objectives. In theory, every objective would state them, but in practicality it is useful to describe overall prerequisites for a given set of objectives. Prerequisites let learners know what they need to know before they can learn, and can serve as guide for a cycle of learning (e.g., in order to take the intermediate class, learners must take the beginning one). When spelled out, they allow the learner to self-assess their abilities. Because some learners over-assess or under-assess themselves, it is important to be detailed and precise.

DETERMINING OUTCOMES

In their simplest form, outcomes are the results that come out of some achievement. But in the case of learning objectives, this means more than simply doing the objectives. It means being able to apply the learning. Learning for its own sake is kind of abstract unless it can be demonstrated in some application.

Outcome-based objectives are common in education, especially in schools where there is some requirement of proof of learning. The outcome is often demonstrated by hands-on lab participation or shown in some other work, such as homework problems or a report. Outside of education, outcomes are harder to require. In work settings, learners often learn the objectives but may not have to demonstrate an outcome (though they may end up applying the objectives at work).

An example of determining outcomes: Let's go back to the browser example, and look at using the bookmark manager. Learners will learn how to save URLs for sites, edit the properties of the URL, and create divisions with folders and rules. Having done this, learners will accomplish the objectives. However, they then need to apply the learning to a real-life setting, most notably the work environment. The outcome may be that a supervisor wants everyone to have the same bookmarks, labeled the same as everybody else, to create uniformity in the department.

> "*Outcome*: Learners will be able to create bookmarks for two work-related sites (NYPL and Internet Library), rename them in a uniform way described by the department, and create a uniformed named folder (Online Library Sites) to put them in."

Why determining outcomes is important: As noted, it is not enough to simply learn for learning's sake. Objectives must be applied. This not only provides reinforcement for the learners, but also gives validation to the learning itself. If nothing else, outcomes should be practiced as part of the learning session in an attempt to replicate a real-life setting.

NINE USES FOR OBJECTIVES

To put things into perspective, here are various uses for objectives:

1. Learners can scan them to get an idea of what is involved in the learning prior to registering or signing up for the learning
2. Learners can review them ahead of time to assess themselves and find out whether they meet prerequisites
3. Learners can use them in a session as an outline to keep track of where the learning started and is going
4. Learners can use them after a learning session for review and reinforcement
5. Instructors can prepare by reviewing them for updating the learning prior to a session
6. Instructors can use them to describe expectations and outcomes, and to select instructional methods, materials, and strategies

7. Instructors can use them as an outline in classes, especially if they are detailed and complete
8. They can be reviewed by someone who needs to approve the learning to get an idea of what is covered in a learning session
9. They can be used to review the person who undertook the learning to ensure that learning took place and can be applied

REVIEW TEMPLATE FOR DESIGNING OBJECTIVES

Scan the following as checklist for comprehension of this chapter on designing objectives. Also, you can use this list as a quick checklist when you are creating objectives for your own technology teaching.

What do you need to know about the participants?
Who are they? Why do they need to undertake this learning?
How and/or where will they use the learning?
What are their general skill levels? How much related experience do they have?
What are their general learning preferences?
What type of learning environment is likely to be used?
What is the comfort level (e.g., with technology in general, hands-on activities, group participation, etc.) of the learners?

What are the specific skills that will be learned?
Which type of skills or behaviors (verbal, intellectual, etc.) does this learning attempt to facilitate?
What are the specific actions to be accomplished?
Which action verbs best describe those actions?
Are the actions demonstrable? How?
How can you make them easily understood by learners?
In what sequential order should they be undertaken?

What circumstances influence the learning?
For each of the objectives determined above, what settings or circumstances influence each one?
What context is needed to understand the learning?
When/where will the learning need to be applied?

To what degree/extent will the skill be carried out?
- For each of the objectives determined above, what degree or extent should be placed on them?
- How does the degree help establish a measurable outcome of the learning?
- Can it be expressed in terms of how much, how long, how far, etc.?
- Is "mastery" implied? If so, can it be explicitly stated? If not, to what extent does the learning extend?

What previous learning is needed?
- Given the objectives above, what specific skills or steps are necessary to be able to achieve them?
- How can prerequisites be stated to let learners know what is required of them?
- Can learners use them to self-assess?
- Do previous prerequisites already exist? Are learners aware of them?

How will the outcomes be described?
- Given the objectives above, what practical outcomes can be established?
- How can the outcomes be applied and measured?
- Do the outcomes match learners' needs?
- Do they fit the learners' environment and context?

What expectations might be encountered?
- Given the objectives above, what expectations might be anticipated?
- Can you identify expectation by talking to learners prior to learning?
- How can expectations be described? How can they be addressed?
- Where/when can the expectations for a specific learning situation be stated so that learners are made aware of the expectation placed on them?

3 STEP 3: DEVELOPMENT

In this chapter you will:

- **Identify a seven-point outline for developing learning plans**
- **Describe outcomes of learner analysis**
- **Determine objectives based on learner needs**
- **Define steps needed to achieve objectives**
- **Identify four strategies for facilitating learning**
- **Match strategies to objectives**
- **Review template for developing learning**

INTRODUCTION

Here we take the results of analysis and design, sort them out and begin developing learning. Analysis scouts the territory, design sketches out the blueprint, and development puts the building together. This is the point where ideas are converted into concrete applications. To develop learning a teacher or trainer must be able to identify needs, find objectives to fill them, and match objectives to strategies to facilitate learning.

This is the phase where we actually build a lesson module or learning plan. Each plan pulls together three main elements: objectives, strategies, and materials to facilitate learning. It is called a plan because it has a framework and structure for how to proceed.

AN OUTLINE FOR DEVELOPING LEARNING PLANS

If design equates to a blueprint, development equates to a manual. It is helpful when developing learning to have an outline for how to pull all the pieces together.

- Interpret learner needs
- Determine outcomes
- Develop measurable objectives
- Identify steps to achieve objectives

- Select strategies to facilitate learning
- Develop supporting materials
- Select exercises to practice
- Identify before/after activities

Below is a brief overview of the steps in the outline. This shows how all the different pieces of analysis and design are integrated. Following that is an in-depth look at what all is involved with each step of the development outline. And, as always, plenty of examples are provided to give you practical insight into what to do at each point.

INTERPRET LEARNER NEEDS

Once you've done learner analysis, you should be able to describe the needs of the learner. You should have identified what they are doing now and where they want to or should be. The gap between the two is the need. You should have done some sort of analysis to determine what training or teaching will help.

DETERMINE OUTCOMES

From the description of needs select specific outcomes that are to be accomplished or achieved. Outcomes are demonstrations that the learning has taken place and fulfills the need. They should be specific. For instance, if your objectives state that learner will be able to find information using an Internet search engine, then an outcome would be actually tracking down the information which matches the need—a Web page, document, table of data, fact, etc.

DEVELOP MEASURABLE OBJECTIVES

Remember, we said that objectives should consist of four parts: participants, behavior, circumstance, and degree. Measurable means that the learning will be demonstrable, either in the activity used as part of the learning or through the outcome, as noted above. It helps to use action verbs for the behavior and define a specific degree, such as "three times" or "until the steps are complete, yielding a correct result."

IDENTIFY STEPS TO ACHIEVE OBJECTIVES

Identifying measurable objectives makes it easier to state the steps needed to achieve them. These steps make up the basic process for the learning involved in an objective. Steps should be listed sequentially to provide a simple, easy-to-follow process (see Figure 3.1). How the steps are arranged may differ depending on

Fig. 3.1: Sample of Identifying Steps

Motor skill: Learner will be able to click and drag a shortcut icon into Recycle Bin	**Intellectual skill:** Learner will be able to insert name into header in Word document
1. Position mouse arrow cursor over the appropriate icon 2. Click once (left button) and hold down 3. Without letting up on button, drag arrow cursor, with icon attached, toward Recycle Bin 4. Drag icon until Recycle Bin is highlighted 5. Release button to deposit icon	1. With document page opened, click on menu View and select Header and Footer 2. Within dotted area designated for the Header, type name 3. Click on Close button to exit Header 4. Click on Print Preview icon to view header as it will be printed 5. As you type text, notice that header will appear the same on every page

Fig. 3.2: Sample Strategies To Facilitate Learning

Motor skill: Learner will be able to click and drag a shortcut icon into Recycle Bin	**Intellectual skill:** Learner will be able to insert name into header in Word document
• Demonstrate: show learner how it is done • Practice: try each part; for instance, first try to grab icon and move it, then try to drag it, then try to deposit it • Have several icons to practice on	• Demonstrate and define terms; for instance, distinguish between editing while header is open versus normal editing • Provide documents for learners to practice on

the type of learning (affective, motor skills, verbal knowledge, intellectual skills, problem solving).

SELECT STRATEGIES TO FACILITATE LEARNING

Strategies are techniques, approaches, and tools used to help convey knowledge or skills to facilitate learning. These can include demos, the use of analogies, discussions, hands-on exercises, etc. (see Figure 3.2). Strategies are also likely to differ according to the type of learning involved.

DEVELOP SUPPORTING MATERIALS

Supporting materials are those things used to assist the strategies. While they may most likely include handouts of various kinds, they might also include a variety of artifacts. For instance, in order to demonstrate e-mail through an analogy to the post office, you might use envelopes and an address book.

SELECT EXERCISES TO PRACTICE

An important part of facilitating learning is to reinforce that learning through exercises. These could include a variety of formats: hands-on exercises, following steps, showing work, etc. Or they could include in-class quizzing, homework, or tests. Exercises are likely to vary depending on the participants. For instance, you probably won't give seniors a test, although you might give them an exercise to complete later as homework.

DESCRIBE OUTCOMES OF LEARNER ANALYSIS

Your analysis has helped define what learners need. Before those needs are described as learning objectives they must have a defined outcome. An outcome is the demonstrable proof that an objective has been achieved. If an objective states that a learner will be able to identify two definitions for the term "Web server," then the outcome would be a two-part definition or list of descriptions. If that sounds too obvious, good. Outcomes should be the natural and obvious extension of the objective.

Unfortunately, too many times people do not think of an outcome as the result or proof of the learning. If the outcome is not stated explicitly and is not measurable, the objective itself may be questionable. For instance, to state that learners should "understand what a Web server is" is not only vague, it is not measurable! A teacher or trainer cannot explain what a Web server is and does, and then simply ask the learner, "Understand?" The nodding of the head that would accompany such a response is not a measurement of the objective. You can't measure understand—you can measure an outcome.

Maybe people don't use outcomes because it takes time and effort. Too often we think of the information-learning model from our past in which we listened to lectures, took notes, and then repeated back memorized facts during tests. That's fine for verbal information, although there are several ways to specify verbal outcomes.

To specify an outcome, ask yourself, "What do I want the learner to learn at this point, and how can I make that demonstrable and measurable?" Sometimes it helps to work backward. Imagine the learners demonstrating this learning—what are they doing? Perhaps they are writing definitions, or matching terms

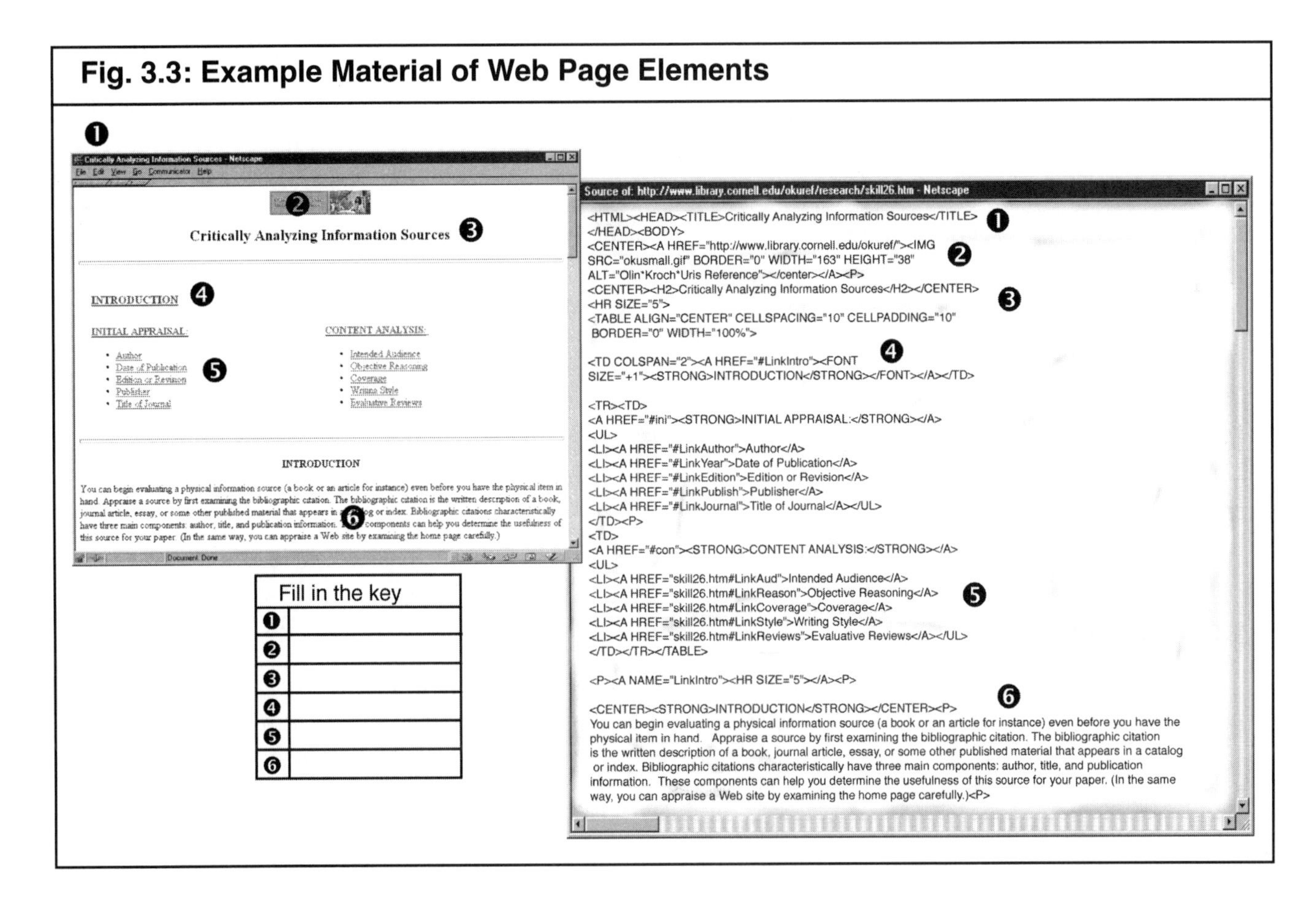

Fig. 3.3: Example Material of Web Page Elements

with explanations. Perhaps they are performing a series of steps using the computer. Perhaps they are finding some specific information (using a search engine), or producing something (using a software application). Whatever it is, it should be describable in a way that is specific and concrete.

As an aside, determining outcomes is useful in more than the development of learning objectives. For almost any decision, or process, there should be a measurable outcome as a result. For instance, a man went to a chiropractor with neck pain. The chiropractor told him she wanted him to use a "neck training device" to help alleviate problems in the curvature of his vertebrae. When asked what the objective of this physical therapy would be, she responded, "to alleviate your pain." He asked her, "What measurable outcomes are we trying to achieve?" She responded, "We will reduce the amount of ibuprofen you take daily from 800 mgs to none, and improve the curvature of your neck by 15 degrees, as shown by x-rays, in six months." Outcomes for any objective should be explicit and measurable.

An example of describing learning outcomes: In its simplest

Fig. 3.4: Sample Exercise To Measure Outcome	
Objective: Freshman students will be able to search in two different search engines and compare 20 results, distinguishing between unique and duplicate results.	
Exercise: Search in two different search engines and compare results	
1. Open search engine: www.google.com 2. Search using this query "customer management" 3. Review the first twenty titles/URLs	1. Open search engine: www.fast.com 2. Search using this query "customer management" 3. Review the first twenty titles/URLs
4. Identify duplicate and unique hits	
Unique (mark with G or F):	Duplicate to both:

form, an outcome can be contained within a learning objective. The objective, "while viewing a Web page, describe the various elements present and how they are arranged on the page," contains the outcome, "describe the elements on a Web page." It is up to the teacher or trainer to decide how the outcome will be performed and measured. Remember, quizzes and tests are possible for students, but not likely to be accepted by older adult learners. Perhaps a print of a Web page could be used upon which elements were circled and labeled (see Figure 3.3). Or perhaps this could be demonstrated generally by discussion, asking participants to speak up and point them out.

Other outcomes may need to be made explicit to be demonstrated. For instance, if the objective is, "perform and compare results of searches on two different search engines," the outcome is the result of performing a series of tasks. But how do you make that explicit and measurable? One way might be to have the learner perform the same search in two search engines and determine how many of the first 20 results of each search overlap, to demonstrate the need for searching multiple search engines. Or the learners could identify which of the first ten results best answered the information need, to demonstrate which of the search engines was better for certain kinds of searchers (see Figure 3.4). The reason for comparing should be stated explicitly.

Why learning outcomes are important: As noted, the outcome is the primary way in which the learning is demonstrated. It isn't enough to say the learning objective should be measurable. The trainer or teacher has to decide how it will be measured. Without a demonstrated outcome, there is no way to determine if the learning has been effective.

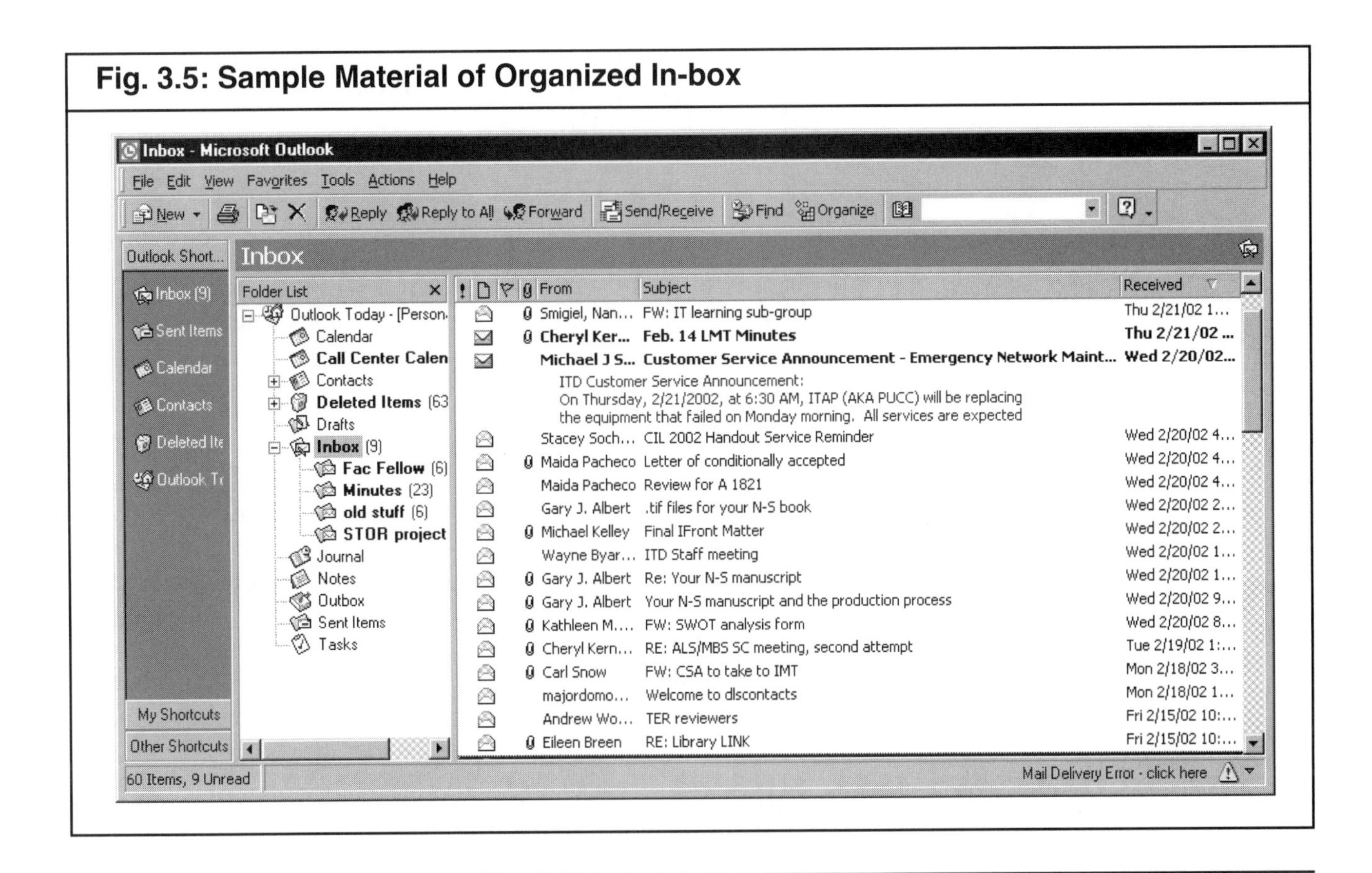

Fig. 3.5: Sample Material of Organized In-box

DETERMINE OBJECTIVES BASED ON LEARNER NEEDS

Learning objectives are dealt with in the preceding chapter, but since this chapter focuses on development, learning objectives will be addressed more specifically here. In particular we want to look at how the learning need links with the objective.

As part of the analysis, you should have broken down general needs into more specific ones. In a previous example we looked at the need for teaching people how to use their e-mail more effectively by training them how to create and use folders for all the messages in the in-box. We break down the general need (organizing e-mail) into more specific ones (creating and using folders).

You should first ask yourself if there is need for an affective learning objective here. As you recall, affective objectives address the emotional or mental well-being of the learner. In this case, learners initially may not buy in to rearranging their in-boxes (it takes extra time and effort, after all). It would be helpful to have an objective that shows them why this learning is useful or important to them. For instance, see Figure 3.5.

> **Need:** Learners should feel that an organized in-box is helpful to their work
> **Objective:** Shown the difference between an unorganized in-box and one organized with folders (circumstance), beginning e-mail users (participants) will list (behavior) benefits, including finding specific messages quickly and easily (degree)
> **Outcome:** A list of benefits

The next thing to do is decide what actions are needed, and what outcomes will be associated with them. Then determine the objective. It is likely that the first reaction is to jump directly to creating folders. But if you think about it, there has to be a reason for initiating this action. It implies that e-mail can be categorized and kept in corresponding folders. Thus the first action is to categorize e-mail, and the outcome is the categories that will then become folder names. So the next objective and outcome become:

> **Need:** To identify topical differences between messages
> **Objective:** When faced with an in-box of three different types of e-mail messages (circumstance), beginning e-mail users (participant) will be able to determine (behavior) three categories to distinguish them (degree)
> **Outcome:** Three categories that can be used for folder names

The objectives from here on should be fairly straightforward. (Note: as stated previously, it may not be necessary to repeat the same participant over and over again if it is clear that these objectives go together as part of an introduction for beginning e-mail users.) Think through each of the objectives as if you were writing a computer program—be precise describing each individual objective's circumstance, behavior and degree:

> **Need:** To ensure that filed messages can be easily retrieved from one place
> **Objective:** Prior to creating folders (circumstance), open (behavior) the appropriate directory folder (degree) in which to create new folders
> **Outcome:** Opened window of the e-mail directory folder
> *(Note: If how to open a folder has not been previously covered, it will be explicitly detailed when the steps for achieving the objective are specified)*

> **Need:** To create e-mail folders when needed
> **Objective:** Having identified names for folders (circumstance), create a new folder (behavior) in the e-mail directory, type (behavior) the name of one of the categories (degree), and repeat for all categories
> **Outcome:** New folders with the names of categories of e-mail

Notice that while creating the objectives you may experience what has been called an "iterative process" in creating learning—the process may be interrupted or go through a cycle more than once before it is finished. For instance, we started out with the idea that we interpreted our needs into two specific objectives, creating and using folders. Then we realized there were affective and preliminary needs to fill. Thus we had to revise our outline to list benefits of organizing e-mail and determining categories before stating the two objectives directly related to creating the folders.

An example of determining objectives based on learner needs: A further example of objectives is detailed at the end of this chapter, based on the previous discussion of introducing users to the Web prior to using Internet search engines. The objectives for the series below are based on the analysis that beginning users of search engines need a conceptual understanding of what a search engine indexes. Specific outcomes include defining and explaining various things that make up the Web (files on the Internet, parts of a Web page) based on the analysis that users are familiar with only a few types of files. Also basic descriptions of how things work (browser-server interaction, links) are included based on the analysis that users did not have a fundamental understanding. How to use a Web browser (as well as mouse, etc.) is covered elsewhere as a prerequisite.

The first objective addresses why the series is needed and useful—this is informational as well as fulfilling an affective need. The next one describes what kinds of things are found on the Internet, to serve the need for explaining what can be found in a search engine. The one after that explains the elements on a Web page to fill the need for later discussion about how those things get indexed. Next is an objective that discusses how things move around on the Web, as a preface for the subsequent objective that explains how links/URLs work.

Why it is important to base objectives on learner needs: As has been pointed out elsewhere, developing objectives in a vacuum can result in useless learning. Disregarding learner needs is tantamount to teaching blindly. And it is not enough to stop at identifying and meeting general needs. Specific needs should be broken down to form useful objectives.

DEFINE STEPS NEEDED TO ACHIEVE OBJECTIVES

Having identified the objective and outcome, we need to ascertain the steps needed to achieve them. Objectives are usually specific enough that they require only a few steps to achieve them. This is a key point in developing the outcome and objectives. In order for them to be measurable, they need to be concise and concrete. This should make defining the steps to achieve them easier.

The reasons for defining steps are:

1. You need to verify all necessary actions for the process
2. It is important to ensure that you have the steps in proper sequence
3. It is critical to identify what prerequisites are needed before someone would begin the process
4. You will use the list of steps as the process for achieving the objective, which will ultimately build your guideline for instructions

The purpose of delineating these steps is to create an outline for the teacher to use in facilitating the learning as well as a procedure for the learner to follow. For the teacher, each step must be identified and defined clearly and easily. Exactly where does the process start, what prerequisites are needed, and how does one proceed? Do not take anything for granted. Steps can (and usually should) be as small as needed to be precise. For the learner, simple step-by-step procedures allow learning to happen easily and stick quickly.

In order for everybody to achieve each of the objectives you've created, they must perform steps required in a process associated with it. You must be able to list the individual steps of a process that will lead to the outcome, which is your objective. This means that for any learning session that has several objectives there will be several processes.

How many steps are in a process? That depends on how complex the objective is, and whether there is more than one process you require. For instance, look at this objective: "After saving several Web site URLs as bookmarks, the learner will be able to open Netscape's Bookmark manager, create folders, rename and delete bookmarks, and move them into folders to eliminate duplicates and organize bookmarks by topic."

In actuality there are several processes involved in this objective:

1. Opening the Bookmark manager
2. Determining topics for grouping bookmarks
3. Creating folders
4. Renaming bookmarks
5. Identifying duplicate bookmarks
6. Deleting bookmarks
7. Moving bookmarks

Each of these processes has steps that the learner must follow to perform the process and work toward the objective. For instance, to open the Netscape Bookmark manager:

1. Move mouse pointer to "Bookmark" menu item
2. Click to select "Edit Bookmark" function
3. Open manager window to edit bookmarks

Learners must have prerequisite skills to get them to the first step, previous behaviors that allow them to take the steps to learn a new skill. These must be identified, then put all together—which of the prerequisites are needed at which point? What must come first? For instance, to use a search engine, doesn't the learner need to know the URL? Identify the basic or advanced search box? Locate tips or help screens? When spelled out, prerequisites serve as a self-assessment tool for learners—can they do these things before they take the training? (If not, they need to learn before coming to class.) In addition, consider the learners who will receive the instruction. For instance, steps you would take or how you would proceed to teach search engines might be different if your learners are librarians, high school students, seniors, or scientists.

Look carefully at an objective and walk yourself slowly through the steps required to achieve it. Pretend you are explaining this to someone who has no knowledge or understanding of the task. Starting at point A, what must you know to proceed to point B? Define the terms you use and describe the steps in concrete nouns and action verbs. Sometimes it helps to move through a sequence of steps backward—review where you are at point C and determine how to get there from point B, and then figure out how you get to point B from point A.

Take for instance, the objective, "Having identified names for folders, create a new folder in the e-mail directory and type the name of one of the categories." Where does the process start? You should have the appropriate directory folder open. What prerequisites are needed? The directory folder was selected and the categories defined. Exactly how do you proceed? Be as precise as possible:

1. Right-click inside the selected directory folder
2. From the shortcut menu, select "New Folder" option
3. When the new folder appears, type the name of the category
4. Hit return to set the name of the new folder
5. Repeat the process for each folder needed

A complete objective with steps to achieve it would look like this:

Objective: Having identified names for folders (circumstance), create a new folder (behavior) in the e-mail directory, and type (behavior) the name of one of the categories (degree).

Prerequisites:

- Appropriate e-mail folder open
- Categories for folders determined

Steps:

1. Right-click inside the selected directory folder
2. From the shortcut menu, select "New Folder" option
3. When the new folder appears, type the name of the category
4. Hit return to set the name of the new folder
5. Repeat the process for each folder needed

Outcome: New folders with the names of categories of e-mail

Note that the prerequisites for this objective are the previous objectives in this particular module. This happens in many situations, and makes sense since usually several objectives are linked together to form a module. Even the first objective of a module will have prerequisites. They will either point to another module, or will require learners to come to the learning with requisite skills. However, be sure to describe them explicitly, especially if the prerequisite skills are from outside the learning module. You must alert learners to where the learning begins so they will know what is required of them.

In the e-mail folder example, there are several prerequisites that must not be assumed or taken for granted. Learners must have "mousing skills," which include being able to open and close windows, right-click, choose menu options, move between windows using the task bar, etc. They must be familiar with the system they are using—they must understand the terminology and names for icons and features (such as folder, directory, task bar, in-box, etc.).

It is important to spell out these prerequisites as clearly and detailed as possible. In fact, prerequisites should be presented to the learner *before* the learning takes place so they know what is expected of them. Learners should be able to use the prerequisites to self-assess if they are ready for the learning. If they are not, they must understand that they should acquire the needed skills before undertaking the learning. If possible, prerequisites should be described with steps to give the learner the most comprehensive picture of what is required. For instance:

Prerequisite: Mousing skills, which include the ability to:

1. Drag mouse pointer to select points on the screen, such as title bar, scroll bar, and task bar
2. Click on minimize/maximize buttons, click in textbox, click at the beginning or end of text to highlight it
3. Drag window by holding mouse button down after selecting title bar and moving it on screen
4. Resize window by moving mouse pointer to a corner or edge where the pointer changes to a double-headed arrow, then holding down mouse button and moving sides of the window

You may be wondering, "Does it have to be this detailed? That's tedious!" The more precise you can be, the better the learning will be. There is no doubt that many teachers and trainers gloss over some of the details of learning. Hopefully you are reading this book because you want to know the best way to develop learning modules. Try to be detailed. Your learners will benefit from your diligence.

Another example of defining steps to achieve objectives: In the example below, the objective on describing elements is basically a verbal learning objective. How do you determine steps for verbal learning? Basically, the same way, by moving step by step through the process that a learner must take.

Objective: While viewing a Web page, describe elements, arrangement, and how things are represented by HTML

Prerequisites: Web page opened for viewing in browser

Steps:

1. Identify icons, links, text, images
2. Identify placement of elements on the page (top, bottom, proximity between) and point out the 250th word on the page—these are prerequisites for a later objective that explains relevancy ranking (weighting based on placement of text on the page) and the fact that some indexes only process the first 250 words
3. View page source to reveal how elements are coded in HTML, pointing out title, URL/links, image file names, and the words on the page (this demonstrates the text that a search engine "spider" actually indexes)

Outcome: Match terms to items on a printout of a screen

Why defining steps is important: Many instructors make the mistake of thinking that if the learner doesn't understand something they can always explain it to them. This is a fallacy for several reasons: instructors may not realize that learners do not understand; learners may be too embarrassed or reluctant to speak up and ask for help; or learning may be happening outside the presumed classroom confines (i.e., on their own).

Learning is based on steps, sometimes called scaffolding. Learners grasp one bit of learning at a time and build from there. Each step in a series is just like a brick in a wall—without the preceding one, the subsequent one will be unstable. Thus, steps must be described in sequence and in detail to be successful.

Fig. 3.6: Ways To Present Information

Breakdown of ways to present information		
Communicating	**Showing**	**Participating**
lecture	demonstration	hands-on exercise
discussion	analogy	written exercise
sharing	show-and-tell	group interaction
question-and-answer	simulation	homework
moderating		practice
counseling		

IDENTIFY STRATEGIES FOR FACILITATING LEARNING

Gagne suggests instruction should be composed of nine steps: gaining attention; explaining objectives; recalling previous learning; presenting new information; guiding learning; performing outcomes; allowing for feedback; doing some kind of evaluation; and providing for reinforcement (Gagne and Briggs, 1974). Gagne's advice can be followed during development of the instruction, and during the implementation of it. Here we will focus on four aspects: developing ways to present information; creating materials to support learning during the event; generating exercises to help measure outcomes; and creating things to do before and after the learning event to prepare for and reinforce the learning. The next chapter will discuss strategies for implementation.

WAYS TO PRESENT INFORMATION

If you think about it, you can probably name several ways to present information for learning, such as giving a lecture, overseeing a hands-on exercise, etc. We'll define several of these below. Once you have the content established through objectives, outcomes, and steps, you need a way of conveying that content. These are the means for delivering content to learners.

It is helpful to categorize them, as shown in Figure 3.6. Communicating basically involves one-way or two-way communication, usually verbally, between the teacher and learner. It relies on a lot of verbal interaction. Presenting by showing usually means the teacher or training uses physical devices to support learning. It relies more on visual interaction. And participating usually implies active involvement in doing on the part of the learner. It relies more on tactile interaction. Together these address three learning preferences—hearing, seeing and doing—since, as we mentioned earlier, it is best to use them in combinations to accommodate all kinds of learning styles and preferences.

This list, though not conclusive, gives some examples of strategies for presenting:

- Lecture: A talk given to present basic ideas and concepts. Considered a necessary way to share verbal information, it is the least effective method for learning. Lectures need to be supplemented by showing and participating strategies. When creating a lecture, be sure to use a format that follows the objectives. For instance, describe what you're about to cover, explain it, and then list the steps to do so and follow with examples. Do not simply throw terms and information at the learner.
- Discussion: An interactive session in which learners react to and talk about a topic. The teacher often leads a discussion with input from learners based on their thoughts, experiences, etc. When using a discussion strategy, be sure to ask for responses by encouraging learners. For instance, ask questions like, "How does that strike you?" or "Why do you think that happens?" and thank them for participating.
- Sharing: Less formal than a discussion, sharing is often used to debrief an exercise or situation. When using sharing as a strategy, be sincere and ask people how they feel about issues or problems. For instance, you might ask, "How does it feel to have this new technology thrown at you?" Listen to their concerns and address the benefits the learning will have for them.
- Question-and-Answer: More structured than a discussion, question-and-answer sessions often seek to elicit specific responses to test outcomes or measure learning. When creating a question-and-answer session, be sure to use wording directly from objectives and outcomes. For instance, rather than ask, "How many people think they can do this now?" you might ask, "Who can name the first of the three steps?"
- Moderating: A teacher facilitates discussion without being a part of it. Often used for affective sessions where it is important that the learners hear responses from each other. When using a moderating strategy, let learners ask each other questions—guide them, but let them use their

own words to describe the experience. For instance, you might ask, "Do you think everyone feels the same way about this?"

- Counseling: Implicit in counseling is a sense of guiding the process of learning by the teacher, not necessarily directing the content, as in a lecture. Counseling works on a one-on-one basis, for reinforcement, remedial help, or affective learning. When used in a session, be sure to keep questions objective. For instance, rather than say, "Why are you upset with this?" you might say, "What might be upsetting about this technology?"
- Demonstration: Showing the steps to achieve an objective. Ironically, a demonstration is not always effective for inductive learners, who try to follow along but get lost glancing down at what they are doing. When creating a demonstration, follow the steps per your objectives. For instance, you might explain a step, show it, then explain the next step, show it, etc. Note when you are about to click on something!
- Analogy: A good analogy is a comparison or contrast of several different aspects between two things. (Not to be confused with a metaphor, which can be vague—such as "the Internet is an Information Superhighway.") When creating an analogy, be sure to use as many comparisons or contrasts as you can. For instance, when comparing e-mail to the postal system, you might contrast composing, addresses, delivery, contents, etc.
- Show-and-Tell: Learners showing or describing an object or system. Useful for debriefing or testing outcomes and measuring learning. When using show-and-tell as a strategy, ask learners to explain specific pieces. For instance, you might ask a learner, after an exercise, to discuss whether it could be done with more or less steps.
- Simulation: Teachers or learners can assume characters in role-playing, or objects can be personified (e.g., "the browser sends a message via the URL and wakes up the server to ask her for the file located at that address"). Best used for concepts. When creating a simulation, think along the same lines as an analogy and compare and contrast between the simulation and real things. For instance, in role-playing how mail is delivered, note the difference in size of information delivered or timeliness of delivery.
- Group interaction: Small group participation, usually in an exercise, and sometimes including debriefing of the experience. When using a group interaction, keep size to four or fewer people and assign a role for each person. For

Fig. 3.7: Learning Types and Formats

Learning types and formats				
Affective	**Motor skills**	**Verbal knowledge**	**Intellectual skills**	**Problem solving**
sharing	demonstrate	lecture	discussion	question/answer
counseling	show-and-tell	analogy	moderating	group interaction
role-playing	practice	written exercise	practice	hands-on exercise

instance, person A could be the recorder, person B could report back, person C could read the directions, and person D could keep track of time.

- Hands-On exercise: Usually implies an additional means to practice after a demonstration. For instance, a lesson could be taught by demonstrating and practicing an aspect, then demonstrating and practicing another aspect, and then using an exercise to practice the two together for reinforcement or to measure outcomes. When creating a hands-on exercise, include problems that result in errors. For instance, when teaching about typos in URLs, include a URL with a typo.
- Written exercise: Often used for testing outcomes, a written exercise could be used in lieu of hands-on exercise. When creating a written exercise, the point is to reinforce the learning, not to make the exercise hard or long. For instance, a multiple-choice test may be more reasonable for adult learners than fill-in-the-blank.
- Practice: Usually implies students following a lecture or demonstration and then practicing the steps shown. When incorporating a practice session, use the format mentioned: demonstrate, then practice, demonstrate, then practice. For instance, show step one, then have the learners try step one.

The trick to successful designing is in matching the various presentation formats to the type of learning at hand. An easy way to do that is to look at the five types of learning: affective, motor skills, verbal knowledge, intellectual skills, and problem solving (see Figure 3.7). While some of the formats work for more than one type, some of them lend themselves more naturally to one than another.

MATERIALS TO SUPPORT LEARNING

For our purposes, materials to support learning refer to learning aids developed to supplement the presentation. These are derived

Fig. 3.8: Sample Lecture Slides (Verbal and Intellectual Objectives)

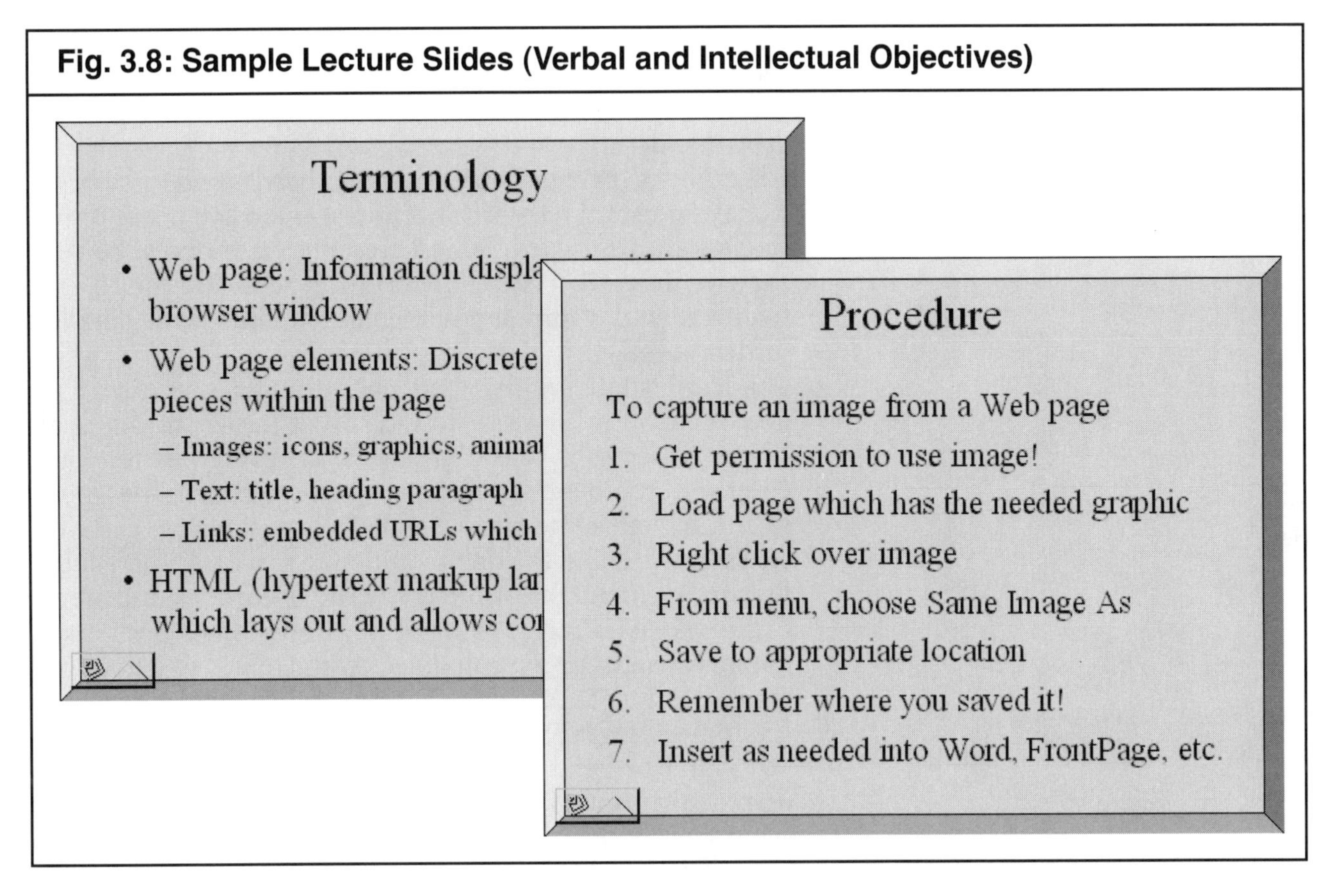

directly from the objectives. This does not include the tools used to help deliver the learning, such as computer or the projected image of the computer screen. The term "materials" may seem to imply print products, but the focus is on the content, not the format (which could be a Web page instead of a printed page).

As with strategies for presenting information, there are several different types of materials that can be developed to support learning. Probably the most often used is the ubiquitous handout. A handout can come in many forms and be used for several reasons—to demonstrate, to remind, to detail, to supplement, etc. They are used to support learning, but are not substitutes for objectives, outcomes, and steps (though they often include the same content). For example, the steps to achieve an objective may be listed along with pictures showing screen shots of a Web browser (Figure 3.8).

This list, though not conclusive, gives some examples of primary supporting materials:

- Lecture Slides/Overheads: The slides or overheads can be useful for people to follow along with, take notes on, or refer to later. When creating lecture slides, try to follow the format of the objectives. For instance, title a slide with the objective and then show one or two steps. Do not crowd slides.

- Class Notes: Because slides and overheads tend to be brief, it may be useful to reproduce more extensive notes to use in class or afterward. When creating class notes, be sure they tie in to the class. For instance, replicate the format from the slides and supplement each slide with detailed information.
- Guides: Usually a "how to" handout, a guide may be needed to show several steps at one time, or to detail all of the aspects of one objective (i.e., prerequisites, outcomes, steps, items used, etc.). When creating guides, keep in mind that some people will follow them in class instead of watching the demo, and many will use them later. Be sure to include a context for the objectives and steps. For instance, a screen shot of a Web browser bookmark manager might be accompanied by the text, "to edit bookmarks, follow these steps," and then list the steps.
- Examples: It is useful to refer to additional examples when you don't or can't refer to them as part of the lecture or demonstration. When using examples, be sure to label them clearly. For instance, if you show a good example and a bad example you must be sure to point out the difference.
- Artifacts: In some cases, it may be necessary to show a process, such as cleaning a mouse ball, and use the actual object. When using an artifact, consider how the learners will react to it. For instance, if showing them how to clean a mouse, you might want to use old ones, rather than the ones attached to the computer (otherwise, what would you do if a mouse ball rolled away and couldn't be found?).
- Manual: More complex and complete than a guide, a manual may serve as a source of constant referral during a detailed session on specific operations. When creating a manual, be sure it follows the format of a guide, but provides clear division of topics. For instance, use section dividers to separate different objectives.
- Supplemental Reading: Adds depth or definition to explanations, and can be taken away for reference or reinforcement. As will be shown later, it is useful to include reading in the event there are mixed learners in a class—basic reading (definitions, fundamentals, etc.) for novices,

in-depth for advanced learners. When including supplemental reading, make sure it is clearly marked so as not to confuse learners with other materials used during the learning session. For instance, you might title a lengthy handout "Tutorial on Advanced Searching."

- Bibliography/Webliography: Where it is impossible or inconvenient to provide supplemental reading, it is useful to supply a list of where the materials can be found. Many people think of a Webliography as not only an online version of a bibliography, but one that points to things online, as opposed to in print. When creating a bibliography, be sure to include where the sources can be found. For instance, note call numbers for books in the library or where full-text articles can be found online.

To develop supporting materials for handouts:

1. Review Objectives: identify which ones may need supporting materials to help explain or to be used in the future
2. For guides, manuals, etc., describe the outcome covered, list steps (definitions, etc.), and provide an illustration (screen shot) if needed
3. Keep the "look-and-feel" of the handout simple and consistent—use a logo or icon as appropriate

Supporting materials are chosen in much the same way as the other strategies—by linking to objectives. Once you have defined your objective you must identify its type in order to determine which materials best support it. For instance, a verbal objective would likely include a list of definitions, and an intellectual objective would likely include the use of a model or workflow diagram of steps. Motor objectives might include a diagram, such as a cross-section of a mouse explaining how it works. Affective objectives might include anecdotes or stories relating successes. Problem-solving objectives might include puzzles or problems, with hints to help work through them.

Most training will employ a straightforward outline for structuring information on a handout. First, be sure to describe the objective and outcome for the procedure or skill. Then describe each step concretely and succinctly. Last, be sure to include graphics wherever possible—screen shots, images, diagrams, etc. These types of materials should be able to stand on their own when the learners refer to them later when they need them. Other types—lists, testimonials, exercises, etc.—are likely to be put down in a format that the instructor thinks best fits the circumstances. A

Fig. 3.9: Handout For "Searching For Journal Articles"

OUR Library

Using Advanced Search

This search is similar to Assisted Search in the Libraries catalog. You can combine sets of search terms as well as select the article type, publication year, and peer-reviewed articles.

Insert Your Screenshot Here

1. Select one or more databases
2. Click the Adv Search button
3. Type in your search terms
4. Select words Anywhere or a specific field
5. Combine your search terms with AND or NOT
6. Select a Source document type
7. Select publication year or range, as appropriate
8. Click the Search button

Searching Options

The following list shows valid options to create search queries. These can be used in all of the search methods in this index.

Boolean operators to combine words/phrases: AND, OR, NOT
    Ex. train NOT rail,
Proximity operators to qualify proximity: IN, NEAR, WITH
    Ex. baseball NEAR stadiums
Wildcards: * for multiple endings of a word, ? for zero or single character
    Ex. comput* (for computer, computing, computable, etc.)
Phrase: automatically searches words as phrase—no special characters
    Ex. New York
Nesting: ( ) use parentheses around groups of words with operators
    Ex. (baseball NEAR stadiums) AND New York

Search Results

Search results can be displayed in 2 ways: Brief Display and Full Display. Brief display provides a list, which includes the article title, source with volume and pages, and date. In addition to the article title, source, and date, the full display includes the author, abstracts, and descriptors.

Insert Your Screenshot Here

Fig. 3.10: Sample Exercises For Searching
Before class...
Bring a topic to search when you come to this class. Choose something about which you will feel comfortable writing a short paragraph which will compare information from four different sources (Web pages or documents).
During class...
1. Write down your topic (e.g., cold remedies) • Identify other, similar words (e.g., flu, sore throat, medicine, recuperation) • Identify terms which are broader topics (e.g., health, seasonal disorders) • Identify terms which are narrower (e.g., cough syrup, throat lozenges, Echinacea) 2. Compose a search query for your topic, and ones for a broader and narrow search 3. Conduct all three searches in both Fast and Google 4. Compare the first 20 results and identify duplicate and unique results
After class
Ask your friends which other search engines they use. Do your searches again, and compare the results to your original searches. Which search engines did you find most useful in finding "good" results? Additional results?

handout with quotes from testimonials along with photographs may be appropriate in one situation, whereas a list of benefits without testimonials may be appropriate in another.

Supporting materials should be as "clean" and simple as possible (see Figure 3.9). Don't overdo it with either graphics or text—let white space be your friend. Don't mix more than two kinds of fonts types and sizes. Make sure images photocopy well. Use colored paper to help distinguish between handouts in class ("please look at the exercise on the green paper"). But try not to overwhelm attendees with handouts—if you have more than two or three, consider putting them in a folder or handing out ones of after the class at the end.

EXERCISES

The purpose of an exercise is to give context and meaning while practicing an objective or demonstrating an outcome. It can be used in addition to simple practice of steps after a demonstration. The difference is that practice usually involves mimicking steps that have been demonstrated, and an exercise involves working through a problem with a context (see Figure 3.10). For instance, practice might be stated, "right-click, select New Folder, and type name" whereas an exercise might be stated, "now identify categories for your e-mail messages and create folders for them."

Exercises can be included during a session as further practice, at the end of a session as reinforcement or evaluation, and after a session for further reinforcement and practice. In the "demo-practice-exercise" approach, after several steps of an objective or objectives have been practiced, a contextual exercise is given to further stimulate learning. Often, an exercise may be held to the end of a session after all objectives have been covered—the intent is to give learners a chance to practice the new learning in a way which emulates their usual work setting. An exercise assigned as "homework" would allow learners to practice after the session, but often requires some kind of proof to demonstrate that the exercise has been completed.

An exercise usually involves hands-on work, although it could be developed as a "paper-and-pencil" exercise. There are actually a few aspects of computer work that are easier to practice without the computer. Examples of this might include writing down category names for e-mail folders or brainstorming broader-than and narrower-than terms for a search query.

BEFORE/AFTER

As you look over the collection of objectives you have developed for a session, it is worthwhile to consider preparation or reinforcement activities that you could assign to learners. Activities for before the learning may include verifying that prerequisites are met, or preparing something to use in the learning session. Activities for after the learning may include reinforcement or a required demonstration of applying the learning.

There are several things you can do before or after a learning session to help make the learning stick. As noted, it would be wise to have participants check prerequisites beforehand. Also, sending them the objectives to prepare them for the learning would be useful. And if there are any exercises requiring thinking up a topic or problem, you might want to ask them to bring it with them. After-the-learning activities might include repetition of practice, additional exercises, or further reading to explain concepts and demonstrate applications. Keep these in mind as you develop supporting materials.

Fig. 3.11: Matching Types of Objectives With Strategies

Affective	Motor skills	Verbal knowledge	Intellectual skills	Problem solving
Ways to present information				
sharing counseling role-playing	demonstrate show-n-tell practice	lecture simulation written exercise	discussion moderating practice	question/answer group interaction hands-on exercise
Materials to support learning				
guides	artifacts	slides/overheads suppl. reading bib/Webliography	class notes manual	examples exercise answers
Exercises				
role-playing	demonstrate	quiz/test	hands-on exercise	hands-on exercise
Before/after				
bibliography	examples	suppl. reading	manual	written exercise

(Table heading: Matching types of objectives with strategies)

MATCH STRATEGIES TO OBJECTIVES

Once you have identified and selected strategies, but prior to developing them, you want to make sure they will be suitable for the learning you are going to facilitate. There are several aspects you might consider, starting with a key question: how will the strategy (or supporting material) help learners? Note that strategies used to present can also be used as exercises, and materials to support can also be used for before and after the learning.

First, will the type of presentation strategy used help facilitate a given type of learning objective? Figure 3.11 presents a list of strategies and types of learning—this table seeks to match presentation strategies to the types of learning it can best help support. For instance, lectures and written exercises deal with verbal information and question-and-answer and group interactions deal with problem solving. Second, will the type of supporting material work better for one type of learning than another?

Third, how can you best match exercises to demonstrate outcomes to types of learning? Motor skills would probably be better demonstrated than quizzed, although an in-class quiz would work for verbal information. Fourth, how can you match activities for before and after the learning? Perhaps by matching background reading for affective-related topics, and an exercise for problem solving.

Finally, remember to take into account the different types of learners. Don't simply provide visuals for visuals—provide a variety of strategies to reinforce learning.

These are suggestions, not necessarily definitive answers—sometimes you have to try variations or combinations to see how they will work (see Figure 3.11).

An example of matching strategies to objectives: Let's go back to our e-mail example from earlier. The third and fourth objectives are intellectual (process-oriented) skills:

> Objective 3: Prior to creating folders (circumstance), open (behavior) the appropriate directory folder (degree) in which to create new folders
>
> Objective 4: Having identified names for folders (circumstance), create a new folder (behavior) in the e-mail directory, type (behavior) the name of one of the categories (degree), and repeat for all categories

Knowing that these steps are process oriented, we might use a discussion presentation strategy (along with some demonstration) to reinforce the process of creating and naming folders. We might use step-by-step guide handouts so that participants have examples to follow. We could have them demonstrate that they can achieve the outcome as we watch them perform it in hands-on setting. And we might give them a handout of the complete process from beginning to end either before the class to familiarize them to the terminology, or after the class so they can refer to it.

Why matching strategies to objectives is important: Some strategies just don't go well with some types of learning. Too often in the past learning has been treated as all verbal information, and learners have been overdosed with lectures and quizzes. To make learning stick, it is important to vary strategies to match different learning objectives (see Figure 3.12).

Fig. 3.12: Sample Learning Outline

Learning Outline for "Introduction to Internet Search Engines"	
Objectives, and steps to achieve them	**Strategies to facilitate learning**
I. What does "Intro. to Searching on the Internet" cover? Given a list of objectives, participants will be able to determine how appropriate/relevant the course is for them 1. Describe relevance A. Share stories/nightmares of frustration from not finding information easily B. Ask what their expectations are regarding class 2. Review objectives by describing what this course will do and help them achieve	Type: Affective and Verbal Outcome: Overall confidence in the course Prereq: Experience using some kind of search engine; familiarity with terms: search engine, searching Strategies: Sharing Materials: List of objectives Exercise: Have each person introduce herself and give a response
II. What can you find on the Internet? Given examples, participants will be able to list seven types of information found on the Internet 1. Show Web pages with the following: graphics, pictures, text, tables, documents, animation, movies 2. Show equivalent files on a floppy disk (show files with extension and size in a directory window)	Type: Verbal Outcome: A list of information types on the Internet Prereq: Familiarity with browser, URLs Strategies: Demonstrate variety by showing sites and compare to files on a floppy disk Materials: Web page URLs Exercise: Q&A to identify items
III. What is a Web Page? When viewing the source code for a Web page, participants will be able to identify three elements: text, links, and images 1. Show a Web page, then File\View\Page Source to show the HTML code for the page 2. Discuss various elements: header info (title) and body (text, links, images) 3. Compare and contrast between page and code	Type: Verbal and Intellectual Outcome: Match information type with its HTML code Prereq: Definition of HTML Strategies: Demonstrate HTML source code & elements and compare Web page and code page Materials: Handout show source code with text, links, and images circled Exercise: Q&A to identify elements
IV. How does a URL link work? When shown a URL for a link in the status line, participants will be able to identify domain name, first level directory, and any html page names, and compare this to a phone number	Type: Verbal Outcome: Description of parts of a URL Prereq: General identification of URL Strategies: Compare and contrast Materials: Whiteboard to write on

Fig. 3.12: Continued	
1. Show URL and describe each part 2. Compare to geographic correlation of phone number (i.e., region of state, area within city)	Exercise: Q&A to reinforce description Before/After: Guide sheet to refer to later
V. How does the Web work? Given an analogy, participants will be able to define browser and server, and describe two functions for each 1. Server: Internet computer with server software A. Keeps track of files and makes them available B. Waits for and fills requests for files 2. Browser: Program which follows links (URLs) A. Uses URL address to make request of server B. Retrieves and displays files on the screen 3. Role play the functions: several people play servers and one person acts as browser following link URL determined by instructor	Type: Verbal and Intellectual Outcome: Description of two functions for both server and browser Prereq: Definition of computer, network; familiarity in selecting files from A:, C:, etc Strategies: Personify/Simulate: server is like file keeper; browser is like a courier or delivery person Materials: Copies of Web pages in folders which are marked with their URLs Exercise: Participants play the role of browser going over to server and requesting a page from a folder

REVIEW TEMPLATE FOR DEVELOPING LEARNING

Scan the following as checklist for comprehension of this chapter on developing learning. Also, you can use this list as a quick checklist when you are developing materials and strategies for your own technology teaching.

Have you identified outcomes?
  Do you have measurable outcomes for each objective?
  Do you have a method for measuring them? (e.g., watching learners perform the task, or having them circle file menu options on a printout)
  Can you demonstrate to learners that the outcomes are relevant?

Did you develop measurable objectives?
  Having identified outcomes, do you need to revise objectives?
  Does every objective have an audience, behavior, circumstance, and degree?
  Are there additional objectives needed (such as affective or motor skills)?

Have you listed prerequisites? Are there means for the learners to gain the prerequisite skills or knowledge (e.g., previous courses, Web sites, etc.)?

Have you listed steps to achieve the objectives?
Does every objective have steps to achieve outcome?
Are they concrete and succinct steps?
Can steps be performed by learners?

Which strategies will you use to facilitate learning?
How will strategies support the objectives?
Have you accounted for presentation, support, and exercises?
Can you match strategies to types of learning?
Will they help to demonstrate the outcome?

REFERENCE

Gagne, R. M., and L. J. Briggs. 1974. *Principles of Instructional Design* (2nd edition). Fort Worth, Texas: Holt, Rinehart and Winston.

4 STEP 4: IMPLEMENTATION

In this chapter you will:

- **Define and address qualities of adult learners**
- **Identify an outline for presenting learning**
- **Distinguish between presentation formats and styles**
- **Describe things to do to prepare for presentation**
- **Identify traits of successful presenters/teachers/trainers**
- **Describe benefits of co-teaching or using assistant**
- **Identify and address problems working with technology**
- **Describe benefits of having a back-up plan**
- **Discuss dealing with problem participants**
- **Review template for implementation**

INTRODUCTION

Now we take the things we have designed and developed and put them to work. First and foremost, the learners are key, not you the implementer. You have to understand their learning perspectives, address their needs, and facilitate their learning. As you implement the learning there are certain aspects that you can manipulate or put your personal spin on to achieve a satisfying result for everybody. Implementation is partly a matter of reading and responding to your audience, and partly a matter of manipulating your style to suit the environment of learning.

QUALITIES OF ADULT LEARNERS

We spoke earlier about learners in the chapter on analysis. Here we want to talk about qualities of learners as they impact implementation. There are some differences between adult learners and what we think of as "school students" (including those in college). Approaches must give attention to how learners react to the presentation. Remember that under the ARCS model, we said we must find ways to gain attention, make learning relevant, help students feel confident with the learning, and satisfy their needs.

During a learning session, an important aspect of how you get the learner's attention relates to how you treat the learner. Don't treat adult learners like children—don't patronize or condescend. Adult learners tend to be self-directed and learn best in structured settings. Be direct—explain clearly why and how the learning applies to them. Make it personally satisfying for them, since this is a great influence on motivation. Try to break sessions into 20-minute units to keep things simple and focused (be sure to recap and reinforce at the end of each section).

To make the learning relevant, it is important to make learners feel their efforts are worthwhile. In addition to checking their needs, make sure they want to learn. For instance, cover the objectives and then ask, "Is everybody comfortable with that?" Whenever possible, bring in their experiences, perhaps literally by asking, "What kind of experience have you had with this?" Use examples and exercise from their day-to-day activities (which you should have picked up from analyzing the learners).

There are certain things you can do to help build learners' confidence in the learning process. Repeat activities often to establish habits and build learning. Build on their experiences and incorporate them into experiential activities. They will build up confidence in their learning quickly and easily when it relates directly to them and involves hands-on exercises. Also remember to incorporate a variety of senses (auditory, tactile, visual) to support learning.

Learners will be satisfied when they are able to see an immediate application of the learning. Wherever possible, ask them to bring problems or questions that they can use in hands-on exercise. For instance, in a search engine class, ask them to bring information needs for which they want to find resources. And they will be satisfied when they feel they are a part of the learning, so be sure to incorporate two-way communication into the session.

An example of accounting for adult learner qualities: When presenting the learning session, make it clear that getting through the objectives is part of their work—by associating the learning with work it takes on a mature feeling as opposed to the "immature" or remedial connotation that teaching might convey. Emphasize that the outcome of the learning is to allow everyone to do his or her job better (or to accomplish technology related efforts, such as working online, easier). Focus on the learners' perspective, for instance describing how it will help them when they are at work or at home performing these steps later.

Why qualities of adult learners are important: The bottom line is that adult learners want to be treated like adults. They don't want you to teach at them, they want you to share. And they want you to understand things from their perspective. That does not necessarily mean that you are intimately familiar with the circumstances behind every learner's situation, experiences, or motives. But it should mean that you are willing to find out, and that you are cognizant that each person is unique.

AN OUTLINE FOR PRESENTING LEARNING

The key elements for presenting information are: review any previous related information, discuss the learning objectives about to take place, present the learning, reinforce it, and conclude. Though teaching should not be conducted like a speech there is some application to the adage, "Tell them what you're going to tell them, tell them, and tell them what you told them." In this case we focus on the learning, not simply on the content of a speech.

It was mentioned earlier that Robert Gagne offered a nine-step approach to presenting a teaching or training module. His wisdom has withheld the test of time, and so it is presented here as an outline to follow.

- Get the learners' attention
- Discuss learning goals/objectives
- Review previous sessions
- Present the content
- Facilitate learning
- Make learners demonstrate learning
- Give feedback
- Assess outcomes
- Reinforce learning (Gagne and Briggs, 1974)

Get the learners' attention by pointing out how pertinent the learning is (see Figure 4.1). This might be done by showing a significant achievement in time or money savings that might be made as a result of the learning. Or you might point out problems people will have without the knowledge or skills, and how gaining them will help.

What do you do if participants have no motivation whatsoever? This often happens in the educational setting when students are required to attend a class for which they think they already

Fig. 4.1: Sample of Presentation Examples

	Presentation for: Troubleshooting Your Computer
Get attention	Show numerous examples of error messages, interspersed with pictures of frustrated people in various poses
Discuss goals	Describe outcomes (decreased frustration and increased productivity) and review specific objectives
Review	Discuss prerequisites and whether participants can meet them
Present	Move through each objective and their associated steps
Facilitate	Demonstrate step, have them practice, etc.
Learner demo	Participants will demonstrate that they can accomplish six exercises
Feedback	Instructor will walk around and give feedback during exercises
Assessment	Participants will review objectives and do self-evaluation of progress
Reinforcement	Students will be asked to send e-mails of subsequent trouble-shooting successes or problems to the instructor

know the learning. First, make it perfectly clear why they are there: their instructor has deemed it necessary. Second, just like any other review of material they already know, ask them to answer the questions, rather than you telling them. If students know how to search the online catalog, make it a simple review, not a lecture. Third, explain to them that the purpose is to ensure that all students know the exact same facts and can demonstrate the exact same skills. Fourth, remind them of outcome. Presumably, they are there to learn how to find a book or journal article because the instructor will require them to do so. Point out that they can use this time to get a head start on the assignment. Fifth, don't worry too much. Instructors often jump to the conclusion that because one student is bored, all students are bored. This isn't usually true. Remember that sometimes you have to aim to satisfy the learning of the majority of learners, not necessarily 100 percent.

Discuss learning goals and objectives to familiarize the learner with what will take place. This overview is an important way to acclimate and orient the learner. It is important to start off on the right foot with a clear idea of what will be covered.

Review information and knowledge from previous sessions or experiences and show how it is connected to the current session. Be sure to highlight or go over prerequisite steps to make it clear where the starting point is. Remind the learner or refresh her or his memory. This prepares the learners and gets them into "learning mode." It also helps to set the pace, initiate and stimulate learning for the rest of the session.

Present the content (through lecture, demonstration, etc.), following objectives and using strategies to facilitate learning and achieve outcomes. This is the part that most instructors are familiar with and is the heart of the learning session. The greatest percentage of time is spent here, as noted in the chapter on development. Additional perspectives on format and style (i.e., how to do it) are given below.

Facilitate learning by pointing out how learners can learn. Sometimes known as meta-learning, it is important that learners realize what they are doing successfully to achieve learning. This can be done in part by encouraging practice and problem solving. It can also be achieved by asking them to reflect on what they just did and why it was effective for them. Sometimes you might "pull back the curtain" and reveal why a particular strategy was used for a specific learning objective.

Make sure learners practice in some way to demonstrate that learning has happened. This is as essential for you (to see that learning has taken place) as it is for them (to reinforce what they have learned). Often this can be achieved by having them demonstrate steps that you have presented, and then put all the steps together in an applied exercise. (This might be called the "demo-practice-demo-practice-exercise" format.)

Give them feedback to help reinforce the learning. This may be as simple as praise for trying, or encouragement to keep practicing. Make sure that you make note of achievement (i.e., "getting it right") as well as overall progress from beginning to end (i.e., success at completing all aspects of the learning).

Use some method to assess and measure that outcomes have been met. As noted, this step is often overlooked. It can be combined with the exercise portion of a session, but it will be incumbent on the instructor to check every person to make sure they have achieved the objectives. Alternatively, participants may be asked to comment on what and how much they have learned, though this is more subjective and not very measurable (unfortunately it is the best you can do in some circumstances). Try to get them to work on something on their own after the learning session and report back to you with their success.

Provide for additional reinforcement through exercises or homework. This is slightly different than learners proving they can accomplish the objectives. It is meant to be a further aid to their learning for the long term. Exercises can include questions for which the answers will be provided later (through e-mail, etc.). Homework might include putting together a list based on the skills needed—for instance, developing a guide to sites for online travel after a class on searching and using Internet search engines.

PRESENTATION FORMATS AND STYLES

Where materials are obvious tools for delivering information in a tangible way, two other elements are important in implementing learning—the format of the session and the style of the presenter facilitating the learning. Format refers to ways to present information. Several of these were mentioned in the previous chapter, and some of them will be discussed further here. Style refers to techniques used by the presenter in delivering content and facilitating learning, including speaking skills and how to interact with participants.

Keep in mind that you undertake a number of roles when you are presenting. Depending on whether you are using a communicating, showing, or participating format, you are respectively a communicator, a demonstrator, or a facilitator. Additionally, throughout the process of implementing a session, you have to wear several hats at various stages: planner, instructor, expert, resource, and co-learner. As the instructor you may need to adopt various styles to help convey learning objectives.

As we mentioned in the chapter on development, ways to present information include the following categories of formats: communicating (lecture, discussion, sharing, question-and-answer, moderating, counseling), showing (demonstration, analogy, show-and-tell, simulation), and participating (hands-on exercise, written exercise, group interaction, homework, practice). Each of these categories has general impacts on implementation.

The communicating format involves one-way or two-way communication, usually verbal, between the teacher/trainer and learners. As such, this format requires strong skills in articulating information. Lectures usually imply one-way verbalization, but it is important to use reflection, question-and-answer, and supporting materials along with them. Keep in mind as you are using any communicating format that emphasizing and repeating main points is critical. Facilitate discussion or moderating sessions with questions that lead to the learning objectives that need to be covered. For instance, to initiate a discussion on types of information found on the Web you might ask, "If a Web server is just like any other computer, what do you think you can find on one?"

The showing format involves visual strategies for conveying learning objectives. Visual representations may use graphics, such as images of connected computers to represent the Internet. Likely, demonstrations will be made showing applications on a screen for all to see. Remember to walk through steps slowly and clearly, explaining concepts and defining elements as you go. Don't assume that everyone has done something similar before, or uses the same terms as you. Be specific and precise—for instance, rather than saying, "Choose that feature from the menu," you might say, "To turn that feature on, move the cursor up to that menu option, click once and hold to pull down the pop-up menu and move the cursor down to select the feature, then release the mouse button to turn it on or off." This way you are more likely to facilitate the learning because you are reinforcing the visual demonstration of steps with auditory commands.

The participating format involves tactile or hands-on participation to achieve learning objectives. It usually supplements the communicating and showing formats. For instance, in the example above you might describe the concept of and need for a given feature, demonstrate it as described, and then ask them to repeat the steps. If the exercise is written instead of hands-on you would ask them to describe, in order, the steps required. Remember that some kind of participation is necessary not only for tactile learners, but to supplement non-tactile learning and provide specific reinforcement.

While format is a product of the environment in which learning takes place, presentation style is related directly to the presenter, teacher, or trainer. Style is sometimes considered synonymously with personality, but in fact refers to skills that can be learned or acquired. There is a difference between being knowledgeable and personable and being a successful presenter. Oftentimes people who are experts and have "good people skills" are fearful of presenting in front of an audience.

Obviously, one of the critical elements needed by a teacher or trainer is a high level of comfort and ease "being on stage." How can you attain this? First of all, you have to be comfortable enough with the knowledge at hand not to be nervous. One thing that leads to nervousness is the fear that something will be left out, or that a question will be asked for which there is no easy or immediate answer. You need to practice your material until it becomes second nature. Or, you need to use a script or outline to ensure that you will cover all the material needed.

If you are nervous because you fear you may omit something or of not being able to answer a question, rest assured that it happens to everyone. Training institutes emphasize that no two sessions are ever the same. If you leave something out, hopefully you will have supplemental materials available for participants to review and pick up anything missed. When a question is asked for which you don't have an answer, either ask if someone else

Fig. 4.2: Checklist For Presenting Skills
Key Questions Checklist for Presenting to Learners
√ Are you comfortable enough with the knowledge to be at ease? If not, cover only those aspects you are familiar with (i.e., limit objectives).
√ Have you practiced your material? If not, start with a script, reduce it to an outline, and try to get to the point where you need only a few prompts (e.g., note cards).
√ Are you on track? As you review the objectives, keep an eye on the clock. If you have to skip anything, be sure to offer to share materials with the students later which cover the topics.
√ Are you comfortable with the audience? If not, try to identify a few people with whom to make eye contact. At the very least, try to project your voice around the room— to the left, right, back, and front.
√ Is your personal pacing at the right speed? Force yourself to slow down by repeating concepts and terms or pausing (i.e., count to five or ten). Don't stand in one place, move around if you can.
√ Are you asking enough questions? Don't ask obvious ones, but do defer to the expertise of the audience. When asked a question, ask if anyone can answer it first.

present knows the answer, or write it down and commit to answering it later.

Some people have a fear of talking to groups. There are various suggestions for overcoming this, such as practicing on family and colleagues, making eye contact with only a few people on opposite sides of the room, or focusing on points in the back of a room. Check resources on speaking and making presentations (Figure 4.2). One thing that is key is for the presenter to remember the participant's perspective. Participants want a presenter who is sincere, knowledgeable, and treats them as an equal. They don't expect the presenter to be flawless or without mistakes—like most people, participants are forgiving and understanding of someone who gets up in front of a crowd to speak. Thus, practice, but know that you are not expected to be perfect.

The ability to talk slowly and clearly is important. Training isn't meant to be a rapid speech, it is a facilitation of learning which takes patience. Make sure people can hear and understand you. Define terms and jargon. Repeat important words or phrases. Face people and stand to get the most volume out of your voice—beware of facing the screen and talking during demonstrations. Speak in short sentences. Pause for emphasis and to let people absorb the learning.

In addition to speaking, it is important to ask questions and listen. There is a school of thought that says that instructors should speak to 60 percent and participants should speak 40 percent of the time. This is based on an assumption that participants bring with them to the learning a variety of experiences and problem-solving skills. Learning is stronger if it comes from multiple sources, including participants. Ask or lead them to stating information that supports learning objectives. For instance, rather than define what URL stands for, ask participants to define it. Allow them to ask questions or make comments—interesting and unexpected insights can come about. Whenever possible, reflect questions back to the group to see if they can work it out. For instance, if someone asks, "What does HTML stand for?" you might ask the individual, "What do you think it stands for?" Or you might ask the group, "Does anyone know what this acronym stands for?" If necessary, call on specific individuals and wait ten seconds for an answer. Be sure to factor time for such participation into a session.

An example of balancing presentation format and style: Imagine that you have a combination of lecture, demo, and hands-on strategies for a session on using Internet search engines. As shown, each of these strategies utilizes slightly different approaches to effectively facilitate learning. During the lecture part it will be important to focus on background and concepts. It will be necessary to explain terms and describe settings, and in doing so that you speak clearly and loudly, repeating key points often. You could describe how search engines are like databases, and ask participants to describe the source, structure, and search-ability of a phone book as an example that you will then compare to *Yahoo!* or *Google*. During the demo part it will be important to show elements of the search engines (help pages, search box, results pages) and steps for using them. It will be necessary to view a projection of these and to precisely describe each step as you move through a procedure slowly. A demo can be led by the instructor, or include participation by asking periodically, "What do you think I need to do next?" During the hands-on portion of the session it will be important to allow the participants to repeat the steps, both to practice and to reinforce what was shown. It will be necessary to move about or check in with them to make sure they are able to accomplish the steps, and you will need to counsel and provide support. Set certain benchmarks for them to achieve—can they focus a search to find fewer hits, or can they use terms found on a Web page to revise their search?

Why working on presentation format and styles is important: By thinking about and crafting both presentation and style strategies you will be able to make the learning interesting, serious, and effective. Often instructors rush into "lecture mode" and think their job is done because they've shared information. Students may nod their heads, but it is unlikely they will have acquired knowledge in a way that will stick with them. By carefully thinking about and using appropriate formats and styles you can enhance learning.

THINGS TO DO TO PREPARE FOR PRESENTATION

It is useful to have a checklist for implementation prior to conducting a learning session. This list should include those things that need to be completed in order to ensure a successful learning session. This list ensures that you have all the pieces ready that you have previously worked on in the design and development phases. It should include:

Learning objectives
Session outline (content, examples, strategies)
Handouts and supplemental materials
Exercises (hands-on practice, "after" activities)
Technology (lab scheduling, hardware/software)
Participants (registration, prerequisites, "before" activities)
Facilities (restroom, break room, etc.)
Back-up plan

Obviously, first and foremost, you should have your learning objectives prepared for a session (see Figure 4.3). Your design should have determined them, and your development filled them out. As part of implementation you want to be sure they are prepared and distributed or made available prior to a session (perhaps online). And you will want to have a brief version of them available for participants to use before and after the session, as well as an extensive version to use as your instructional outline.

Presumably you will have some kind of handouts, guides, or examples to use during the session as well. Handouts describing terms or concepts will be useful to refer to during the class. Guides listing step-by-step procedures or URLs for Web sites will be used during hands-on practice or exercises. Hand these out when you need to use them during the session.

Exercises are also designed and developed prior to the session, but you need to determine how they will be conducted as part of implementation. You need to decide when they will be undertaken during the session and how much time will be allocated for them. Remember you may need to try them out first, or allow extra

Fig. 4.3: Checklist For Presentation Facilities and Materials

Checklist prior to class

What	For/By Whom	When
√ Course: Troubleshooting	Instructor: DSB	March 17, 2002
√ Objectives	• Posted on Web • E-mailed to students	• Prior to registration • Prior to class
√ Instructor Outline	Developer: JS	2 weeks prior to class
√ Exercises	• Developed & tested • Photocopied	• 1 week prior to class • 24 hours prior to class
√ Classroom • Software • Hardware	• Scheduled Ref Desk • IT Department • Room monitor	• Beginning of semester • 2 weeks prior to class • 1 week prior to class
√ Registration		
√ Back-up Plan	• Cancellation ahead of time, if possible • Can do Ex. 1, 4-5 and review handouts	• immediately before class, by phone • or reschedule later (get back to them by e-mail)

time for unforeseen circumstances. You will need to decide whether people will work in pairs or small groups, and how you will check to see that the exercise accomplishes its intent. Also, as noted under technology below, you will need to check whether the exercises can be accomplished on the computers used for that purpose. Otherwise, you will need to make sure any exercises intended as "after activities" are in a format that people will be able to read, understand, and follow later.

You will need to check several aspects related to the technology you are going to use. First, ensure that it is available. Likely you will have to reserve the space or computer room for your session. Then check the equipment—both hardware and software—on the instructor workstation and the participant machines. Make sure all applications are loaded, options you are covering are available, and that exercises can be completed. Last, make sure you have a back-up plan, including placing needed files on floppy or Zip disks in case the network is slow or not available.

You should include your participants on a preparation checklist. If they are required to register, there should be some mechanization for doing so (for example, see Figure 4.4). At a minimum

Fig. 4.4: Sample Registration Spreadsheet

Course	Name	Address	Phone	E-mail	Prereq?	Repeat?
TRBL-01	Downing, Ed	Bldg. 12	8-2984	edowning@	Y	-
TRBL-01	Kingsley, Susan	Green	8-3707	kings@	Y	-
TRBL-01	Long, Wm.	Bldg. 4	9-2255	billlong@	N	-
TRBL-01	Noordstrom, Tom	Bldg. 12	8-0131	noords@	Y	Y
TRBL-01	Shepheard, Barb	Green	9-0046	brs@	N	-
TRBL-01	Yu, Soo	Bldg, 9	8-1181	syu@	Y	Y

you will want to contact your potential participants to ensure they have seen prerequisites and objectives for the session. You may also want to contact them ahead of time in the event there are "before activities" or they need to bring a topic or questions with them. Registration can also be used as an attendance list.

Depending on the length, type, or location of your session, you will also need to check out facilities ahead of time. You need to know if the area will be unlocked or secured, well lit or dark enough for projection, heated or air conditioned, quiet or noisy, and whether there will be interruptions for various traffic. You should be able to tell participants where the nearest restroom is located. And if the session is long, you might want to point out where any break room, etc., is located. This usually presumes that you will incorporate a break during the session, which is something else to consider.

Back-up plan: As noted in more detail below, you should always have a back-up plan in place prior to a learning session.

An example of preparing for a presentation: An additional word about preparation—it is wise to get to a classroom 15–30 minutes early. Unless the space is under your direct control, you never really know how the room or setup will be left until you get there. And you should have your materials prepared a day ahead of time. Always make a few more than you need—you will be able to accommodate any last-minute registrations and give out extra handouts if requested.

Why preparing for a presentation is important: More than anything else, having a preparation checklist is another way of systematizing your instruction. Even if you have been doing training or teaching for many years, it helps to keep a checklist handy as a reminder that Murphy's Law is always lurking around the corner.

TRAITS OF SUCCESSFUL PRESENTERS/ TEACHERS/TRAINERS

There are traits that successful teachers and trainers possess. These are personality characteristics that contribute to style, mentioned above. These are not simply qualities that make some presenters better than others, they are useful elements that aid in implementing teaching and training. While some people are born with such traits, it is important that presenters try to acquire them to help learners buy into the learning.

Charisma is hard to define, but it implies that a presenter tries to be charming in a way that makes the learning interesting. Although you do not want to be so charismatic that you are the focus of the session at the cost of the learning! Qualities of charisma include: making eye contact and listening intently to others; appreciating the humor and insight of others; being extroverted and engaging; giving off an aura of confidence and assertiveness. However, try to avoid some of the negative connotations of charisma: egotism, self-indulgence, being the center of attention.

Empathy implies seeing things from another person's point of view. This is critical because the learner's perspective is important. Remember, you want to know where learners are coming from so that you can make the learning easier for them. Try to make them feel comfortable when presenting their point of view, never make fun of them. Be open and willing to see things from their perspective, and perhaps even doing things their way. If you have to correct them, do so gently and in a positive way by saying, "Try it this way" instead of "No, no, that's wrong."

Humor can help make the learning pleasant. The point is not to tell jokes, but to keep the proceedings light, friendly, and enjoyable. A few jokes or humorous stories won't hurt, as long as they are quick and to the point and related to the learning. Where possible inject humor into the session but don't make light of the learning or intentionally hurt feelings. Rather than make fun of others, sometimes it helps to be self-effacing and make fun of yourself. For instance, you might use examples of when you goofed up to warn others what to avoid.

While praise, or the ability to give it, may not be a trait in and of itself, it is certainly something that successful presenters incorporate into their sessions. Be willing to reward people and their positive behavior with kind words and encouragement. Cheer them on, especially when you can tell they've made a breakthrough or success. Encourage them, for instance with statements like, "You've made it through the hardest part, now we've got just a little left." Thank people for their insights and participation, on both an individual and group level.

Another ability of a good presenter is anticipating problems or questions. This is partly a matter of being so experienced with teaching a topic (or with the participants) that you have a feel for what might go wrong or be misinterpreted. But it is also a matter of constantly checking in with the participants, watching for signs of frustrations or bewilderment. When training with people I know I'll sometimes remark, "You look like you're making a face, did I go too fast?" By catching problems as they happen you can keep the presentation on course and running smoothly.

An example of successful presenter traits: I have long been complimented for my easy-going demeanor while presenting. "You make people feel at ease in situations that could quickly become frustrating," I've been told. And I've asked participants what qualities they see in me as a presenter that makes them feel at ease. The number one comment that participants cite is that the presenter should make everyone in the class feel equal. Too many times I've heard of presenters who talked above a class or talked down to the participants. My approach is to let people know that I'm just like they are: we're all overworked and faced with constant change, and my job is to discover ways to make things easier, and that I want to share these with them. Thus, rather than tell someone they are having problems because they are "doing it the wrong way," I'll tell them that I understand the software is hard to use because I had to learn it from scratch too. "But," I'll point out, "once I learned these concepts I was able to figure out the steps to take to work through this application."

Why presenter traits are important: As noted, it seems some presenters are born with those qualities that make them successful and enjoyable teachers. But all presenters must work at creating an effective presentation, and their own personality is a part of that (see Figure 4.5). Incorporating (but not forcing!) elements of charisma, empathy, and humor can make the learning environment enjoyable and thus add to the overall experience in a positive way.

Fig. 4.5: Sample Presentation Approach: "Practice-Demo"

Example of "demo-practice-demo-practice-exercise"...

1. First step, Instructor: Shows entire process of how to delete a shortcut icon
2. Second step,
    a. Instructor: shows how to "grab" an icon
    b. Students: "grab" an icon
3. Third step,
    a. Instructor: shows how to "drag icon"
    b. Students: "drag icon"
4. Fourth step, Students: practice dragging icons into Recycle Bin

BENEFITS OF CO-TEACHING OR USING ASSISTANT

Any time you present in an environment where there are computers for hands-on practice, you should consider having a co-presenter or assistant. Logistically it is nearly impossible for the presenter to help participants and teach them at the same time. To do so usually results in other learners becoming bored while all attention is focused on one person, and eats into the time available. While a co-presenter usually implies someone who shares teaching responsibilities, it is possible that such a presenter could attend to participant problems. An assistant, on the other hand, implies someone who is entirely dedicated to attending to problems with participants.

An assistant can help by doing several things. Positioned in the back of the room, an assistant can check to see that everyone is on the same or "right" page. This is especially helpful when the network is slow and some machines are less responsive than others. Usually the assistant is not there to make sure no one checks e-mail—don't be surprised if participants are "multi-tasking" between e-mail and the search engine demonstrated.

Preferably the assistant would be familiar with the application being taught. An assistant can help solve or anticipate problems. It is helpful to introduce the assistant so that learners will understand why she or he is there. It is useful to have students raise their hands whenever they are stuck or need to ask questions. The assistant should be able to distinguish individual questions from general ones and relay them to the presenter. For instance, if the assistant notices that several people are having trouble logging in at a particular point, she or he might ask the instructor on behalf of the whole class, "Just to clarify, did you say to leave the password blank at this point?"

An experienced assistant will look for and recognize signs that people are having problems, getting frustrated, or need help. Body language, such as slumped shoulders or repetition of the same motion over and over again are likely clues. Since assistants can see the participants' screens and keyboards where the presenter cannot, they should jump in to aid individuals whenever needed.

An example of using an assistant: Often when I teach a new course that involves using software to develop an application, I insist on using an assistant. For instance, teaching how to use a program to build Web pages involves a lot of detailed hands-on work. Utilizing the "demo-practice-demo-practice-exercise" format helps manage the class, but when the program interface is new or complex, an assistant is required. The assistant should have more than just passing familiarity with the program. Standing in the back of the room, she should be watching all workstation monitors and the participants closely for clues that someone is stuck or getting frustrated. When the instructor checks in with the class by asking, "Okay, is everyone at this point?" the assistant can verify and give the "thumbs up sign" to indicate it is all right to proceed.

Why using an assistant is important: As noted, an assistant serves two roles: she provides quick assistance to someone who is temporarily stuck, and prevents the teacher or trainer from grossly interrupting the flow of the instruction. Both are important to keep things running smoothly and on time. Sometimes you might think the effort is not economically feasible, but if you compare the time of two instructors in a classroom for two hours versus either instructor working one-on-one with up to 20 students for an hour apiece you'll see there is a considerable cost savings.

PROBLEMS WORKING WITH TECHNOLOGY

Since our focus is on teaching technology, it is worthwhile to consider the problems related to technology that may happen during a presentation. Problems can happen with technology used for presentation (computer projector, instructor workstation, etc.), or that used by participants for hands-on practice. Such problems might be with the hardware or the software used. Where possible, be sure to check in advance with the systems department (or equivalent) regarding when and where you are teaching to confirm that all systems are going to be up and functional.

To start with, if it is the instructor's equipment that has a problem, it will affect all participants. Thus, it should be of foremost concern. Be sure to check out the system prior to use, whenever possible. Run through as many demonstrations and exercises as you can. This is especially important when you are at someone else's mercy. Similar installations of software are not always exactly the same on all machines—some drivers or options that are installed on the computer on your desk may not be installed on all computers used for training.

If you are using the Internet, load Web pages in advance of a session and then bookmark them for easy access. By pulling them up on the screen, the pages are also loaded into a temporary memory file on the hard drive called the cache. The cache allows the browser to work more efficiently by pulling up the pages off the hard drive instead of off the Internet. If you lose your Internet connection, you should still be able to view the pages.

If at all possible, check (or have someone else check) every computer that will be used. If it is not feasible to do so, do a check of random computers in the room. Don't assume that participants' machines will work the same as the instructor station—often the computer used by the instructor gets all the attention to the detriment of others.

If you have problems with the technology, don't panic. Keep in mind time management for the session. For instance, if someone complains that his machine just isn't working correctly, switch him quickly to another machine rather than try to fix it. Or consider asking that person to sit with another person and take turns during practice or exercises. If the instructor workstation isn't working correctly, determine immediately which objectives you can cover without doing a demonstration, and ask the class if they are willing to adjust. Remember, you might have to cover only verbal and some intellectual learning objectives, and reschedule.

If you can anticipate potential problems, keep a list of "simple fixes" to remind you or participants of things to do for immediate resolution of problems. For instance, if a mouse, monitor, or keyboard isn't working, make sure they are securely plugged into the computer. If a toolbar isn't showing, have participants look under the "View" menu to highlight and display the appropriate one. If a computer "freezes," try pressing the CTRL, ALT, and

DEL keys to restart the machine. For problems that take too much time, such as installing missing drivers or downloading needed plug-ins, write off the machine for the time being, but be sure to record the problem so someone can fix it for the next time.

BENEFITS OF HAVING A BACK-UP PLAN

You know Murphy's Law—"Anything that can go wrong will go wrong"—and that the warning was invented before the Internet came along! The Internet, or technology in general, can be a trainer's nightmare because things can change or become unusable right before your (and the participants') eyes. It is helpful to anticipate problems that may occur and plan for contingencies. And although a back-up plan is often presumed to relate only to technology, it can also apply to changing gears regarding format or style as well.

Obviously, if the power goes out, you'll have to cancel a learning session (especially if you intended using computers). If just the instructor station has problems, you'll have to decide whether you can walk people through steps without being able to demonstrate them. If not, cancel the class. Otherwise, you'll need to slow down, and determine whether some objectives will not be covered or covered elsewhere. If so, you'll need to have supplemental materials with you. This contingency should be designed and developed as part of the overall training. If the participants' computers stop working or don't work like they need to, then a sudden and major overhaul of your training session is needed. Decide if it is feasible, or even worthwhile, to cover a few objectives, such as verbal or intellectual/conceptual ones.

If you are using the Internet you might consider downloading pages ahead of time. Make sure you get permission in advance—many Web page authors will allow you to download their pages if you promise to use them for one specific date and time (and it doesn't hurt to tell them that theirs are the best examples you could find on the Web). You can do this one of two ways. By pressing the ALT and PRNT SCRN keys, you can capture a window screen and paste into Word or PowerPoint, etc. where you can resize it as needed. (Note: this bitmap capture can increase the size of a file dramatically.) Alternatively, you can open the page in Netscape and choose the Edit Page feature, or open it in Internet Explorer and use the Import Wizard feature. Both of these methods will allow you to save a page to a disk that can be opened later in a browser for viewing.

Regarding the need for a back-up plan for format or style changes, consider changing your approach if the attitude of the participants is dramatic. This depends on not only participants' currents needs, but their current mental state as well. If they appear tired or quite bored, you may need to drop verbal objectives and lectures and do more hands-on or group work to stimulate them. Plan for several options—what to do if they are tired (increase activity), if the content is overwhelmingly over- or undersufficient (revise objectives and take it up or down a notch), or if they are argumentative or resistant to learning (continually focus on the relevance to them).

An example of back-up plans: When drawing up your instructor outline, highlight objectives that require technology and which ones don't. Thus, at any point, when faced with hardware or software failure, a quick reassessment easily can be made. Likewise, note where objectives could be divided if the session had to be canceled midway through the session. If need be, literally write down what you would say under various circumstances, such as when faced with software inconsistencies or a lethargic or resistant group—that way you'll be able to speak calmly and coolly and keep the session well managed.

Why having a back-up plan is important: Murphy's Law (and he never saw the Internet!)—needless to say more.

DEALING WITH PROBLEM PARTICIPANTS

As someone who does a lot of train-the-trainer workshops, I get asked questions about how to deal with "problem participants" all the time. This includes people who have either over- or underassessed themselves for the class and are bored or frustrated, but are very vocal about it. It may also include people who are demanding or interruptive, people who want you to do everything for them, and technophobes.

First, try to see things from their point of view. This not to say you have to agree with or kowtow to them, just that it is helpful to take their mental models or perspectives into account. It may be that you can come up with an explanation for a concept or an analogy that really helps them understand. For instance, if someone pouted that Internet search engines were just too foreign and sat there with crossed-arms, I would respond to them with a quick analogy to point out that they are just like online catalogs or indexes that point to documents or articles instead of books in the library.

Second, be firm. Remind people of the objectives (which should have been available prior to the session). Point out that this is the only thing you are prepared to cover at this time, and if it is too advanced or simple for them, they have three options: 1) sit there and see if there is anything new they might learn; 2) volunteer to help or receive help from others; or 3) leave. Note, some trainers report success matching advanced and novice participant together, but sometimes people do not like the responsibility of helping or to be put in the position of accepting such help.

Be frank. Explain why it is that you can't cover other materials or assist individuals as demanded. Objectives are part of a system of training, whereby you allocate and commit your time to that material you have planned to cover. Time-wise, you can't afford to cover additional material. Likewise, if there is only one of you, you won't be able to help all individuals who have hardware or software problems. That would be disruptive to the rest of the class, and eat into the time already set aside. Besides, in addition to not having enough time to figure out problems with computer installation or setup, it is not your area of responsibility to do so. You may not have the knowledge or permission to do so.

Be sympathetic. Apologize for not being able to make it an equal learning experience for everyone. If a computer is unavailable, suggest sharing a machine and taking turns. Point out that much of the information can be gained by watching and practicing later. Note when the prerequisite or next level class is scheduled, and suggest they sign up.

An example of dealing with problem participants: The most embarrassing time I've ever had was when a portion of the participants were forceful in insisting that I cover material completely different than the objectives set up for the workshop. I held fast to my commitment to cover only those objectives listed. I apologized, but kept repeating that the objectives were fixed and that we could negotiate a future time to discuss their needs. I have to admit I felt quite uncomfortable when they questioned my judgment and expertise. And I was very flustered when several of them walked out midway through the session! I was comforted, however, in the numerous responses from those left in the session that I had handled the situation well, under the circumstances. And I firmly believe that it is entirely inappropriate to redo the objectives once a session has been planned. (Hopefully it is reassuring to know that this is the only "bad" session I have ever had.)

Why dealing with problem participants is important: Remember that you have a limited amount of time to meet the needs of all the participants in a session. Problem participants start out as people whose needs are not being met. And, if for whatever reason you cannot meet their needs satisfactorily, then the session is probably not for them and they should consider leaving. Do not give in to renegotiating content during a session—note that you will be glad to discuss needs later. Avoid trying to appease every single person's needs if it is at the expense of the larger group. Every once in a while you'll end up displeasing one or two individuals, but remember they are a very small percentage of your overall success (and you can't please everyone all the time).

REVIEW TEMPLATE FOR IMPLEMENTATION

Scan the following as checklist for comprehension of this chapter on implementation. Also, you can use this list as a quick checklist when you are implementing your own technology teaching.

How will you define and address qualities of adult learners?
- Have you taken into account attention, relevance, confidence, and satisfaction?
- Are you using examples that draw on their experiences?
- Are you keeping it simple, short, and to the point?
- Have you identified exercises that apply to their situations?

Have you accounted for all steps in your outline for presenting learning?
- Have you included a review or prerequisites or previous related learning?
- Are you prompting them to think about how they are learning?
- Do you have benchmarks to gauge their success when practicing?
- What kind of feedback are you able to provide?
- Have you included exercises/homework for reinforcement?

Which presentation formats and styles will you use?
- Have you considered differences in the formats used for various objectives?
- Are you able to distinguish when you are filling different roles as instructor?
- How will you make sure you feel comfortable in front of a group?
- Do you have a response prepared in case you cannot answer a question?

Have you included opportunities to practice the 60 percent/40 percent rule of talking?

What things do you need to do to prepare for the presentation?
Does your checklist include all materials you need to prepare?
Have you reserved the technology needed and notified relevant systems people?
Have you contacted participants ahead of time (registration, "before" activities)?
Are you familiar with various facilities that might be used?

Have you incorporated traits of successful presenters?
In what ways will you try to be charismatic? Empathetic?
What kinds of humor have you included into the session?
Are you prepared to give praise and positive feedback?

Will you co-teach or use an assistant?
Have you determined if an assistant is needed for the session?
Is the assistant familiar with the objectives and instructor outline?

Have you identified possible problems working with technology?
Have you "previewed" the session to make sure needed technology works?
Have you bookmarked and loaded Web pages as needed?
Do you have a "quick fix" list handy?

Do you have a back-up plan?
Have you prioritized or marked objectives for contingencies?
Are you giving yourself enough time to download Web pages as needed?

Do you have a plan for dealing with problem participants?
Are you ready to be sympathetic and frank, but firm?
Are you able to suggest alternatives for participants who have either over- or under-assessed themselves?

REFERENCE

Gagne, R. M., and L. J. Briggs. 1974. *Principles of Instructional Design* (2nd edition). Fort Worth, Texas: Holt, Rinehart and Winston.

5 STEP 5: EVALUATION

In this chapter you will:

- **Define what evaluation is**
- **Describe why, when, and where to evaluate**
- **Identify what to evaluate**
- **Determine how to evaluate**
- **Identify ways to evaluate learning outcomes**
- **Review a template for evaluation**

INTRODUCTION

The last part of the instruction process is to look at how well we've done and how we can improve. This applies to both the learner and for the instructor. Evaluation has been referred to earlier as an iterative process—you constantly go back to review and change things based on new information or data that comes up as part of the analysis, design, development, or implementation process. That iterative aspect reflects the flexible and fluid nature of the overall process. But there are times when you want to approach evaluation with an analytical mind, to gather information and review it to synthesize ideas for improvement or to develop justification for continuation or changes.

You might ask why you need to evaluate if you spent all the time on analysis and design? The answer is that you should always work to improve instruction. And once you have finished teaching a session, you need proof that it was effective and learning has happened. If you simply teach and then walk away, you'll never know what impact the teaching had on the learners. You won't know if they've learned if you don't ask. Evaluation gives us proof to examine, not simply an affirmative or negative answer. We have to use evaluations to ask which things taught were more helpful and effective. We evaluate both the instruction, and how learning outcomes are applied.

Fig. 5.1: Sample of Formative and Summative Evaluations

Formative Evaluation

Instructor asks student to review draft of materials used to support a group of objectives for a lesson. Student sits in front of computer with the objectives, their outcomes and steps. Next to each set of objectives are questions for the student to answer. Is this easy to understand? Can you follow the steps? Can you achieve the outcome? Instructor sits nearby but does not interact. If student has questions, the instructor simply asks her or him to do however much she or he can with the materials at hand. Afterward, the instructor asks the student if there were any problems, difficulties, confusions, etc.

Summative Evaluation

Having sat through the class, the student is given an evaluation form that lists the objectives covered in the course, and asked to rank her or his level of confidence at having achieved the outcome. For instance:

I feel:	Very Strongly	Strongly	Neutral	Somewhat Unable	Unable
I can delete a shortcut icon	___	√	___	___	___

DEFINITION OF EVALUATION

Earlier it was mentioned that assessment is a process of collecting data or information and analyzing it, with the intent to find out something that will help us design instruction. In a generic sense, evaluation is the process of making a decision or critical judgment based on assessment. For our purposes we use the term *evaluation* to mean a method for making decisions after measuring the effect something has or had. Most people believe that analysis comes at the beginning of our process of developing instruction, and evaluation comes at the end, though it is not always that simple. Analysis may have a little evaluation element to it, and evaluation may have a little analysis aspect to it.

As noted, measurement implies data collection and analysis, similar to what we have discussed in the chapter on analysis. In fact, the methods used in the analysis will apply to and be useful for undertaking an evaluation. The evaluation may involve formal or informal surveys to gather data, but almost always involves direct contact with the participants or instructor and direct observation and review of instructional strategies and materials.

It is noteworthy to make a distinction between what is called a formative versus a summative evaluation (see Figure 5.1). Formative refers to evaluation that goes on while instruction is being

Fig. 5.2: Sample of Generic Evaluation			
Place an X on the image that best reflects your attitude...			
Materials covered were useful	☺	😐	☹
The instructor was satisfactory	☺	😐	☹
I would take another course like this	☺	😐	☹

formed, during the design and development process. This is important to consider because you may save time and effort if you catch problems or difficulties early and correct them before teaching or training. Summative refers to evaluation that takes place after design and development and often focuses on the outcomes or effects the teaching or training has had. Some people use summative evaluations of instruction to make decisions on performance, raises, and promotion. For our purpose we will focus on evaluation as it applies to reviewing and revising instruction after it has been designed, developed, and implemented. Then we'll talk about evaluation of how outcomes are applied.

WHY, WHEN, AND WHERE TO EVALUATE

It does no good to say "Every instruction needs evaluation" if you don't know why you want to do the evaluation. Evaluation uses analysis to make a critical decision about something related to the instruction. Presumably that means to improve it and make it better for the students. But you need to identify exactly what you want to evaluate. Perhaps you find there is not enough time or too much information, thus you might evaluate the instruction on time and amount of content. Perhaps you notice that students are bored and need to check whether the level of materials is appropriate or relevant.

Often evaluations are somewhat misguided. Perhaps you have seen instructors hand out a survey after a class or workshop asking what might be called "generic" questions (see Figure 5.2). For instance, it might ask participants if the session met their expectations, or if they felt comfortable with the topics presented.

But what will you, as the instructor, do when you find out that a majority of people felt uncomfortable and didn't have their expectations met? You won't know what to review or revise. You need to be quite specific. (Even worse, in my opinion, is the survey that asks you to circle a smiling or frowning face to evaluate the session!) The reasons for evaluating may change over time, but you should always identify data that can be analyzed to help make an informed decision.

Another, perhaps not so obvious, reason to evaluate is to report findings to another person or group. It is very rare in this day and age for someone to undertake any training program without being asked to report how it went. Often, prior to an undertaking you will be asked, "What are the outcomes and how will you measure their impact?" Thus, when you gather information it may be for reporting reasons, as well as review and revision. Luckily, the learning outcomes, which are part of the objectives, specify what is to be measured ultimately for impact. For instance, if the outcome says that learners "will be able to revise a search query," then the impact will be measured by asking the learners to quantify how revision helped them (e.g., saved time, frustration, etc.).

Basically there are three times when you might perform evaluation: before, during, or after the session. Usually an evaluation done before a session is part of the iterative process of development. You might bring a potential participant in to verify that your assumptions match reality. Don't simply ask, "What do you think?" Ask specific questions ("Does that description make sense?") or watch without comment as she or he tries to work through the steps of an exercise (if she or he remarks, "I don't know what to do next," simply respond, "What do you think seems like the thing to do?"). You can fill out your evaluation ahead of time, using the outcomes of your objectives as the question. For instance, if participants are supposed to follow steps to revise a search query, your evaluation sheet might say, "Participant could revise query and get: A. no results; B. same results; C. different but 'worse' results; D. different but 'better' results." This evaluation might indicate that you need to revise your instructions on revising a query, or that another search engine should be used in which to perform the revised search.

Sometimes you might do an evaluation during a session, often to gauge the level of teaching itself, or to check the progress or exercises of students in action. Usually, if you are evaluating the teaching, there are two ways to assess it: either by inviting a peer to sit in on the class or by videotaping and reviewing it yourself. In either case, be sure you have a checklist of what you want to

Fig. 5.3: Sample E-Mail Follow-Up Evaluation

Date: Thu, 24 Jan 2002 09:23:34 -0500
From: "SCOTT BRANDT" <techman@purdue.edu>
To: "Kathleen M. Kielar" <katman@purdue.edu>
Subject: Project Management Course, Dec. 10, 2002

Hi Kat,

I wanted to follow up with you on the outcomes covered in the Project Management course you took. You said you were looking forward to using techniques the next day to start on an office move. Would you take a few minutes and answer some questions about which outcomes you've been able to use? This information will help me a lot.

Please answer on a scale of 1 (very much) to 5 (not at all):

1 - 2 - 3 - 4 - 5

1. Were you able to establish project outcomes?
2. Did you identify tools, community, rules and division of labor?
3. Did you specify tasks and steps needed to accomplish them?
4. Did you establish sequence and timeline with milestones?

Let me know as soon as you can, okay?

Thanks a lot!
Scott

evaluate. For instance, you might list physical attributes such as "uses loud voice," or "makes eye contact," or "moves around." Or you might list style attributes, such as "uses reflection" or "pauses after asking a question." Sometimes you will want to check how students are doing during exercises—you might want to verify that they have enough time to do them, that they are not too simple, etc. It is most likely that you would have someone else sit in to observe this, since usually you are too busy facilitating and working with specific questions to be catching everything that is going on.

But most often evaluation is done after the session. Commonly, participants are asked to comment on a session and their responses will be used to evaluate aspects of the instruction. For instance, after asking participants to "Check off the objectives covered in this session which were applicable to your needs," an instructor might learn that it is unnecessary to cover certain concepts or background information for some groups. Evaluation will be based on an analysis of data collected before a critical judgment is made. For instance, if one person checks that ten objectives were pertinent, one person checks that only five were and 12 people check that nine out of ten were, your analysis will have to take into account attitudes and level of the participants. In such a scenario, it is likely your evaluation will determine that the nine objectives are appropriate.

You might also ask participants to reflect a few weeks later on the usefulness of the session, effectiveness of homework, or applicability of supplemental materials and "after" activities (see Figure 5.3). However, it may be that rather than ask participants, you might ask teachers, supervisors, parents, etc., to assess how well learning is applied "on the job" (at home, work, or school). For instance, you might ask a supervisor, "Were your staff able to perform the new activities easily with little review?"

As to where the evaluation should take place, be careful not to be too intrusive. If you ask participants to fill out evaluations during the class, it cuts into time available for learning. If a peer reviewer is used in the back of the room, introduce the peer as someone getting ideas for another class—students may react differently when they know they are being watched and evaluated. Evaluation is best handled at the end of class, but be sure to leave time for this if done onsite. You might consider offering a "reward," such as candy or certificates when evaluations are handed in—otherwise participants may dash off. If you use one-on-one or group evaluations, do so in private to ensure confidentiality and comfort of the participants. And remember that it is preferable to ask them soon after a session, if not immediately afterward. In fact, you might hand them a list of questions to consider after you make an appointment to meet to help prepare their frame of mind for evaluation.

WHAT YOU NEED TO EVALUATE

The goal is to ask about specific aspects that are important to improve the teaching or training. That doesn't mean a level of specificity where you ask if the participants prefer Times New Roman font over Comic Sans MS! Some of the things that you might concentrate on include content, outcomes, and student or teacher performance. It might also include affective questions, which ask whether the participants felt comfortable about specific aspects, depending on the context of the situation (see Figure 5.4 for examples).

For convenience, we'll use the word *content* to refer to the learning objectives, strategies, or materials used in a session. To evaluate the learning objectives, ask specifically if there were too few/

Fig. 5.4: Sample Attitude/Affective Evaluation Questions

Please indicate your attitude:	Liked much	Liked a little	Neutral, no opinion	Not liked much	Not liked at all
Amount of work involved					
Working in groups					
Talking in front of class					
Time of day of the class					
Taking time away from work					

many for the topic, whether they were treated in enough depth, and whether they were applicable. For instance, you might ask, "On a scale of one (low) to ten (high), please rate whether each of the following objectives was treated with enough depth." This may indicate the appropriateness of the level of previous knowledge of learners. To evaluate the strategies used, ask specific questions about the format and style of presentation. For instance, you might ask participants to respond with "Very much; Quite a bit; Very little; or Not at all" to a statement like, "I benefited hearing other perspectives during the Q&A portion of the session." Likewise, materials used during or after the session could be evaluated similarly. You might ask learners to rate the usefulness of the materials on a scale of one (low) to five (high).

It is very important to evaluate if the outcomes of learning are being met. Remember, as part of developing objectives we articulate the outcomes as well. Thus we have a list of outcomes that can be evaluated. Basically this amounts to checking whether learning can be applied in class by completing exercises or outside of the class through activities, homework, or on-the-job performance. Class exercise could be evaluated at the end of a session, asking questions like, "On a scale of one (low) to five (high), how useful were the exercises worked on during the session?" You could ask generally, or about each exercise specifically, "Please rank the usefulness of the class exercises on a scale of one (low) to five (high)." Evaluations for applications of outcomes would have to allow some time for the learner to practice the learning in another setting. A follow-up survey might be handed out at the end of a class, or sent later, to be returned in a couple of months. It might ask, "Please rank, on a scale of one (low) to five (high), how well you feel you are now able to apply the following objectives." And, where appropriate, it may be sent to parents, teachers, supervisors, etc., for third-party input.

Another area of evaluation is student performance in general. As noted earlier, it is hard for the instructor to teach and try to gather data empirically by watching the learners at the same time. Yet there are many things an instructor can learn from their performance. It will be important to measure the learners' attitudes with the end result of making learning more effective. For instance, if learners are bored or frustrated, is it because the level of material is inappropriate or because you haven't found the best way to gain their attention? This could be measured by an attitudinal evaluation question, but be sure to be specific, rather than simply ask, "Were you bored?" because an answer to that question doesn't give you any insight into what to do differently. Ask quantitative questions such as, "Did you feel that the level of material was too high, too low, or just right to keep your attention?" Another way to gauge student performance is to identify what educators call "standard deviation from the norm." This is a scientific indicator of how far from average students score in test. You can use a simplified version of this by having learners record their answers or responses to exercises during a session. Later, you can review these sheets to get an idea of whether the learning was appropriate for students—if everybody got all right answers, the exercise might be too easy, and if they got all wrong answers, it might be too hard.

Likewise, it is important to measure teacher performance as well. In other words, is the instructor effective at implementing learning as well as developing it? As noted, this is probably best gauged by a peer sitting in on the class, but might be done subjectively by videotaping the session and reviewing it later (which is a good idea to do at least once in order to see what you look like teaching). First, you have to identify all the characteristics that you think should be integrated into the teaching: speaking in loud voice; asking questions; allowing others to speak; repeating concepts and phrases; moving slowly through demonstrations; etc. Then have a reviewer evaluate your performance for each characteristic, based on a scale of one to five or one to ten. Review the chapter on implementation, particularly the sections on format, style, and presenter traits for more examples of what to evaluate.

Lastly, we should say a few words about evaluating affective behavior. A lot of surveys ask general questions about overall satisfaction, such as "On a scale of one (low) to five (high), what is your overall satisfaction with this course?" Remember, you won't know why learners scored it high or low unless you ask them to describe why (however, statistics show that few people take the time to write out answers). If you are going to ask them how they feel about the session, be specific: "On a scale of one (low) to five (high) rate each of the factors below that contributed to your overall satisfaction: A. Depth of topic; B. Type of exercises; C. Q&A participation; D. Number of supplemental materials." That way you have concrete data to help make decisions for review and revision.

TECHNIQUES FOR EVALUATING

The basic approach to developing an evaluation is: decide what you want to evaluate; determine what data is wanted and how to collect it; proceed with analysis (see the chapter on analysis); and make critical judgments (revise instruction or report findings). What you want to evaluate depends on how the information will be used. To revise learning sessions, ask specific questions about objectives and strategies. To report impact, ask specific questions about how learners apply the learning. For instance, you might ask, "What will you be able to do with the time saved applying these new techniques?"

To determine the data you want to collect, list the various areas that need to be evaluated. For instance, for a session on basic file management: under strategies you might list demos, materials, and exercises; and under outcomes you might list increase in organization, decrease in disk space used, and reduction of time spent searching for files. For each of these areas, determine how to best ask the question to get the type of answer most useful for evaluation. For instance, rather than ask, "Were the demos good?" you might ask, "Circle all the ways in which the demos were helpful: A. Applicable; B. Varied; C. In-depth; D. Easy to follow." Or, rather than ask, "Did the session give you ideas for saving time?" ask, "Circle each of the following that you feel will help you save time: A. Creating folders for job tasks; B. Using long file names; C. Viewing files in list format to show date created."

Notice that one of the most heavily used means of doing evaluation is to use a scale to rank or rate items or characteristics. Wherever possible, avoid simple "Yes/No" or "True/False" responses and use a scale to give you a more accurate idea of where things fall. The most common scale to use is a five-point one from low to high (see Figure 5.5). This technique is often used in statistical packages or spreadsheets to track overall impact by num-

Fig. 5.5: Sample Lo-Hi Scale Used in Evaluation

Circle the number that best reflects how strongly you feel						
Evaluation statement	**Low**				**High**	**N/A**
The objectives were pertinent	**1**	**2**	**3**	**4**	**5**	**X**
The depth of content was good	**1**	**2**	**3**	**4**	**5**	**X**
The time allotted was just right	**1**	**2**	**3**	**4**	**5**	**X**

ber (i.e., a tally of numbers gives an overall score, such as 23 out of 25, or 92 percent).

In addition to the five-point scale, it is important to include the option of not answering a question, either because the characteristic doesn't apply to that session or the evaluator feels neutral about it. Statistically, an unanswered question is usually withdrawn from the overall total. For instance, if six questions are worth five points each, but someone only answered five of them, the response could be counted as 23 out of 30. However, if the unanswered question is simply thrown out, the result would be 23 out of 25.

A variety of scales can be employed to rate or rank responses. You could use a scale of satisfaction (Very Satisfied, Somewhat Satisfied, No Reaction, Somewhat Dissatisfied, Completely Dissatisfied); you could use a scale of frequency (Always, Often, Not Sure, Seldom, Never); or you could use a scale of feeling (Strongly Believe, Somewhat Believe, Have No Opinion, Somewhat Disbelieve, Strongly Disbelieve). While you can and should use more than one type of scale, be careful about mixing them. Do so only as necessary. Keep sections with different scales separated to avoid confusion.

In addition to the items you identify for an evaluation survey, it is helpful to elicit comments from participants. Comments often give additional insight into the participants' reaction to or feelings about training. However, it seems human nature to avoid filling in long responses, so encourage participants to be brief. Some evaluations ask open-ended questions like, "What impressed you most about today's session?" Others are more direct and ask for specific information, such as "List any objectives, exercises, or materials that were not useful to you." Positive feedback is nice, but data helps you analyze and evaluate.

As noted in the chapter on analysis, surveys can be formal or informal, short or long. A formal survey is printed ahead of time, usually with a variety of types of questions. Keep in mind that the formal survey often looks like a quiz to participants, so try to make it different than other materials used during the session (use different colored or sized paper, interesting graphics, large letters, and easy-to-follow instructions). Informal surveys can be done by asking for a show of hands, or having participants take a blank sheet of paper and write some responses. Short surveys are probably best used at the end of a session to ensure that the evaluation doesn't take up too much time or extend beyond the scheduled time. Long surveys are best used for follow-up evaluations. These could be filled out by either the participants or whoever oversees the application of their learning.

Most participant evaluations are individual, although sometimes you might want a group response. Individual evaluations usually involve a survey, and the aim is to get many of them for a more objective perspective. But an individual evaluation could include a one-on-one debriefing, especially identifying someone who was bored or frustrated and asking them specific questions to find out why. Small group evaluations usually involve an interviewing style of asking questions to gain insight into reactions to the learning. It is best to use someone other than the instructor to ensure participants feel free to respond. A useful technique is to ask everyone to think of things that should be kept or changed relating to various aspects of the session, such as objectives, materials, presenter's style, etc.

Peer evaluation would use similar parameters mentioned above for surveys, only it would be done by a colleague who sits in on the session to observe. This evaluation likely would be longer and more specific than a participant evaluation. A similar survey might be used in self-evaluation, where you review your own work. You could try to remember how or what you did, such as how well you met expected time frame, what kinds of questions were asked by participants, where you had to elaborate or use other examples, etc. But it is likely you will need to capture the session on video and use the evaluation to review it. Again, it would probably be a long, specific checklist.

Teachers and trainers always seem to face a problem getting participants to fill out evaluations. It is generally agreed that for the most returns, have them fill them out responses in the classroom—the return rate for evaluations dwindles dramatically when taken out of the classroom. As noted, you might offer a reward to help encourage participants. You might also consider whether to use a paper or Web form. Paper is convenient, as long as everybody has a pencil or pen with them. A survey has shown that participants tend to fill out Web forms more thoroughly, especially when a script pops up to point out when a question has been skipped. Interestingly, participants are more likely to fill in comments, perhaps because typing responses is easier or more fun than writing them out.

EVALUATING LEARNING OUTCOMES

Probably the first thing that comes to mind when you think about evaluating whether learning outcomes can be applied is to test the participants. And this is a good idea, whenever it's appropriate. Unfortunately, there are many times when it isn't feasible or appropriate to do so. But try to do whatever you can do to check, gauge, or measure the impact of learning.

A test, quiz, or homework assignment is designed to verify whether learners can apply the learning covered in class. The thing about creating fully developed objectives is that you have already started the foundation for a test. For instance, look at an objective such as: "Having identified names for folders (circumstance), create a new folder (behavior) in the e-mail directory and type (behavior) the name of one of the categories (degree)." A question on a hands-on test would ask them to create the folder—proof of it could be measured by looking over the learner's shoulder in class. Or it could be submitted by inserting a screen capture (ALT, PRNT SCRN) of the finished product into a document to be sent by e-mail. A question might ask learners to list the steps needed to perform a procedure. Or the steps could be listed in jumbled order and the learner would be asked to put them into the correct sequential order.

Unfortunately, tests or homework can't always be used in all instruction. Staff may feel they are being treated condescendingly, and there is no authority to ensure that patrons turn in homework. However, if training is directly linked to performance, it may be possible to ask staff or their supervisors to turn in samples of their work to help measure the effectiveness of the training. In other cases it may be possible to encourage learners to submit similar samples of applied learning by rewarding them. For instance, instructors could develop and provide special resources, such as lists of subject specific URLs, to learners who participate in a long-term evaluation project. (Or, as noted earlier, when in doubt offer chocolate.)

REVIEW TEMPLATE FOR EVALUATION

Scan the following as checklist for comprehension of this chapter on evaluation. Also, you can use this list as a quick checklist when you are working through evaluation for your own technology teaching.

Have you considered why, when, and where to evaluate?
- Have you determined whether the evaluation will be for improving the instruction or for reporting on impact?
- Will you use pre-evaluation to verify assumptions before instruction is given?
- Will you be able to impose on students during instruction?
- Is it better to perform evaluations right after a session, or later?
- Do you have a quiet place where one-on-one or group evaluations can be given?

What specifically have you identified to evaluate?
- Do you have specific questions to ask about the content (learning objectives, strategies, or materials)?
- How will you evaluate outcomes?
- Are there aspects of student performance that should be assessed?
- How will you get evaluation feedback on instructor performance?

Which techniques will you use to evaluate?
- Have you reviewed all areas to evaluate and formed questions to evaluate?
- Which type of scale will you use?
- Will your evaluation be formal or informal? Long or short? Group or individual?
- On which aspects would you like students to respond with comments?
- Will you use pencil-and-paper or Web forms to gather evaluation data? What will you use to analyze it (e.g., spreadsheets, databases, etc.)?
- Will you offer a reward or incentive to fill them out?

Can you identify ways to evaluate learning outcomes?
- Are you able to give a quiz, homework, or test to learners?
- Can you ask for a sample of their work?
- Have you extracted evaluations from outcomes?
- Will you offer a reward or incentive to get them to participate?

PART II

BUILDING EFFECTIVE TECHNOLOGY TRAINING PROGRAMS

6 BUILDING A TECHNOLOGY TEACHING PROGRAM

In this chapter you will:

- **Define what a technology teaching program is**
- **Describe who should be involved in building it**
- **Identify factors for an "environmental review"**
- **Define and prioritize teaching/training needs**
- **Determine budget requirements**
- **Describe techniques to achieve sign-off and buy-in**
- **Review template for building a technology teaching program**

INTRODUCTION

The steps necessary to design and implement teaching and training have been detailed. But how do you create a teaching or training program from scratch? Or, as is more often the case, how do you shape one that seems to be growing on its own? As has been emphasized throughout this book, it helps to have a structured plan. Why is it needed?

A TECHNOLOGY TEACHING PROGRAM

Let's look at the difference between a library technology plan and a library's technology teaching program. A technology plan is a document that usually describes the library's philosophy towards providing access to information via technology, as well as its plans for assessing, budgeting for, and maintaining technology equipment and services. A technology teaching program is a working document or set of practices that defines the library's attitude toward supporting and facilitating technology related learning. The two are quite distinct, although they can and should go hand in hand since you will likely need to teach people how to use the technology you support.

A technology teaching program may be known by several names. In the very broadest sense, such a program is used to define and meet your educational needs, and might be part of a staff development or continuing education program. It may be part of the library systems department's support responsibilities, and called a technology training program. Both of these imply that such a program is directed only toward staff. But it could be part of the information literacy program, since technology literacy is often a precursor to programs that facilitate critical thinking. It could also be part of an outreach program, aimed at helping patrons and customers become savvy users of information technology.

Creating a technology teaching program implies that you can define your general needs and describe what you are going to do to meet them. Review the chapter on analysis for approaches to determine what to survey and how to collect data and information. Keep in mind that a comprehensive program might cover all aspects of technology, ranging from operating system mechanics (file management, setting preferences, etc.) to applications (using word processing, spreadsheets, database management, etc.) to library resources (using the online catalog, indexes, etc.) to Internet/Web (using browsers, creating Web pages, using search engines, etc.).

What should it look like when you are done? It shouldn't just be a document, it should be a system; a well-structured program of procedures and practices. For instance, you might start with an overall vision of what technology teaching encompasses or tries to achieve. You might build a program overseen by the staff development department that works closely with the reference and systems staff to determine and develop courses, schedule and track attendance and evaluations, and helps tie the program into the library's performance or promotion system. The program would include not only the courses, but also the coordination and interaction of the people involved.

The vision of a technology teaching program could be simple or detailed. It might merely say, "The library seeks to empower its users and staff to use technology to access data and information." Or it might enumerate the many ways in which it does that, such as, "Provides self-help materials such as how-to guides and online tutorials in the use of library applications, and offers ongoing courses for the public and staff to support their use of library-related technologies." This statement might precede a list of objectives that the program sought to accomplish, such as

"Learners will be able to search for and find useful information resources on the Internet using library facilities."

How and where the teaching program is used depends on a variety of factors that have to be considered for each library's situation. For instance, in an academic library setting bibliographic information or information literacy programs probably already exist—technology could be attached to the program already in place, or made a separate set of courses that are closely coordinated with other initiatives. In a public library setting, it may depend on who does the teaching, the department under which responsibilities fall (such as Reference, or the area in which the teaching takes place [Young Adults]). As is always the case, consideration must be given to how much time and resources are available. It may be that someone in the cataloging department has the expertise to teach in the reference area. Perhaps the course would be designed by one group, developed by a second, and taught by a third.

Another distinction that could be made is that between programs for staff versus those for end users. What's the difference? Some argue that courses developed for end users should be held for staff so that they know what the end users will be doing. For instance, a course on searching the Internet would be of use for staff that can then reinforce the objectives taught in the session whenever they are dealing with end users. Often you will have a more in-depth focus if you are teaching staff how to use technology for their jobs. But it is possible that some course developed for staff might be of interest for end users. For instance, a session on downloading records from the online catalog might be of interest to power users.

INVOLVEMENT IN BUILDING A PLAN

One of the first things you have to do when developing a program is find out who should be included, and to what degree they should participate. Anyone who is attuned to politics knows that you have to touch bases with and include the right people. Check into the politics of the environment in which you work. If you don't have a political bone in your body, you need to think twice before proceeding. You definitely need the support and approval of others. If it's not your role to take the lead, perhaps you could talk with or join the group or team who has input into or control over teaching and training. Maybe you could work with the staff development or systems department person who is most appropriately positioned to suggest and build such a program.

If you can work with a team it can be a blessing when it comes time to actually doing the work. A single developer not only has to do all the design and development, but all the touching bases and letting others know what's going on falls into her or his lap as well. A team can be broken down into several roles: someone to liaison with administration, board of directors, or management; someone to represent the perspective of staff or end users; someone to help with design, development, or implementation; etc.

Think about other areas where there might be barriers to overcome. Will doing something different be looked upon suspiciously, or will there be some who are resistant to changes? Identify who else you might need to talk with or get ideas from. Perhaps people in acquisitions can comment on new indexes coming, people from reference can share frustrated user perspectives, and people from systems can identify new upgrades or applications that would affect teaching or training.

It is possible that people who should be involved may be reluctant. You may need to demonstrate or convince people of the benefits of teaching and training. Or it may be that some people are willing to get involved in some aspects instead of others. For instance, some may like to teach, but others only develop materials. Look for ways to reward people for their involvement. Some rewards include time-off for teaching evenings, trading less desirable duties for teaching, celebration lunches or dinners, or monetary rewards (although this obviously has to be approved by administration). If nothing else, a write-up and picture in a newsletter may be rewarding to some.

FACTORS IN AN "ENVIRONMENTAL REVIEW"

To formulate a vision, you first need to review the environment to get a good idea of what it is you're doing now. Make sure you contact representatives of all areas and groups in the library. Take a look at all the technology used for the library-related services. Some people call this an environmental scan or technology assessment. Each library is different, so don't look at what others are doing until you've taken a good long look in the mirror.

Start at the very top. Look at the library's mission and goals.

Fig. 6.1: Sample Data From Technology Scan

Unit	Workstations	Added new software FY 01	Added new hardware FY 01	# calls to Helpdesk	# new staff	# new projects
Public	44	2	22	58	0	12
Reference	12	9	5	39	2	8
Circ	5	4	5	41	4	5
Spec Coll	2	2	1	20	0	6
Admin	6	1	0	11	1	4
TS	15	14	8	36	5	15
ITD	9	8	6	11	2	25

The best justification for offering new services is demonstration of how they fit and further the library's aims. If the mission is to "provide access to information," then training how to use technology directly supports and facilitates the ability to access information easily and successfully. If the library's goal is "provide expertise in using networked services," then training staff to increase their skills and knowledge directly furthers that goal. It can be measured by the number of staff trained, as well as the number of instances those staff were able to help patrons and customers. Try to determine general categories of training that meet the library's missions and goals, for instance, "staff training in using Web browsers," or "end user instruction in using search engines." (Note: sometimes it is better to use a generic phrase like, "finding, using, and evaluating information in a networked environment.")

Next you'll need to move to the more practical and applied side of the environment. Take a look at the technologies you have. Look at the hardware and software and try to identify the various types—indexes, Web sites, applications (such as browsers, e-mail, word processing, etc.), Internet, Windows, etc. Identify who uses them, how they are used, and how often (see Figure 6.1 for an example). Find out if you can view server logs of interactions with various resources to gather data on numbers or types of users. Inquire about what kinds of problems are encountered using them. The chapter on analysis describes how surveys could be taken of staff or users to identify needs. In lieu of an actual survey, anecdotal evidence is helpful. Several individual accounts of frustration and constant requests for assistance often portray the general picture.

As with the review of mission and goals, you will want to determine possible instruction applications to teach. Not everything need be focused on the Internet—proficiency in using Windows and managing files are fundamental skills that come in handy whenever using computers. However, these areas of need can be broken down further to eventually create specialized classes. For instance, courses on Windows could be broken down into classes on basics (learning where programs reside on the hard drive), file management (creating and organizing folders of information), and customization (changing preferences, options, etc.). Courses on "Searching the Internet" could be divided into classes on Web browsers (managing bookmarks/favorites, setting preferences/ options, downloading plug-ins), search engines (types of, how to use, etc.), and evaluating information. Each of these aims to make staff or end users more efficient and effective users of technology and information.

Last, be sure to review previous plans or practices. Find out what people have done in the past (both successes and failures). Check to see how other programs got started. Look at special "one time" courses that have been offered in the past. Identify why they were successes or failures. Look at any and all teaching or training throughout the system. Determine who is (or has been) involved and use these people as resources with whom you can brainstorm ideas.

PRIORITIZING TEACHING/TRAINING NEEDS

To be successful in starting a program, you will need to prioritize needs and focus on the ones that will have the most impact. For instance, is it better to start with staff, or end users? Don't assume that you can do both at the same time.

How do you figure out priorities? Ask! It may be that your priority will be set for you by edict or consensus. For instance, the board of trustees may demand that training be given sudden and full attention. Even so, get as much input as you can. Remember, there are at least three ways to survey: indirectly, directly but informally, and formally. If you cannot do a formal survey, find out informally what problems staff are having or they see end users having. Think of your self as a scientist—you can't make a very sound conclusion when you have only two opinions. The more input you have, the better decision you can make (see Figure 6.2 for an example).

Fig. 6.2: Prioritization of Needs

Prioritizing need for Advanced Internet Searching course by group

Rank	Group needing training
3	Patrons, Students, and Teachers
2	Patrons, Business, and Organizations
4	Patrons and Seniors
1	Library Staff, Public Services, and Reference
5	Library Staff, Tech Services, Circ, Spec Coll, and IT

The need for input must be emphasized. No matter how well you think you know the situation, additional perspectives add depth and numbers to your cause. By asking others for opinions, input, or to help set priorities you will make your case stronger. Letting others know what's going on now or might happen in the future is the best way to prevent surprise and disappointment. Whenever in doubt, ask.

Don't assume that everything identified as a need can be offered. Part of prioritizing is to determine which things are "mission critical"—which may have to be determined by the administration or board of trustees. In some cases you may have to determine whether the needs are primary or secondary, or even tertiary. For instance, training staff to use Windows NT may be critical because all the computers run on it. However, training them how to use MS Excel for spreadsheet capabilities might be of secondary importance. And training them to use Adobe PhotoShop to design graphics may be tertiary. Likewise, when teaching end users how to search the Internet, it might be critical to teach them how to use a primary search engine (*Alta Vista*, *Google*, etc.), but of secondary importance to teach them how to use a subject directory (*About*, *Yahoo!*, etc.) and tertiary concern to use specialized resources (*411*, *Lycos PeopleFinder*, etc.).

Present this information in a clear and usable way. Create a list of the priorities and develop a timeline or possible schedule for meeting the needs. Put your list in order of importance and priority, and show linkages between needs or courses. For instance, it is useful to demonstrate that skill in using Windows or a browser directly impacts using the online catalog or other libraries resources (indexes, Web pages, etc.). Illustrate a possible scenario of how courses could be offered over a given period of time. For example, show how the course in using a browser could be a prerequisite to a general Internet course that then could be a prerequisite to a search engine course. Demonstrate how much could be done in a specific period of time.

Remember to emphasize those things that support the library's mission best. Focus on the outcomes and impacts that teaching and training will have. For instance, simply teaching how to use search engines might seem to have little connection to the library. However, searching for book reviews should have a direct impact. Likewise, teaching how to use a browser in and of itself seems to offer little for the library's mission. But if the focus of using the browser is shifted to creating a personalized folder of bookmarks or favorites to library Web pages, the impact is more obvious.

BUDGET REQUIREMENTS

If you don't have a budget, it doesn't mean you can't teach or train. But don't fool yourself into thinking you don't need a budget. At the very least, it takes time to do something, and time is money. If you want to teach or train, you have to figure out what other task you're going to stop doing to free up time and resources to perform a new service. For the most part the costs are going to be for time to do the design, development, and implementation, and thus mostly for personnel costs. There will also be costs such as materials, equipment, scheduling, etc. See Figure 6.3 for a sample budget.

Even if you don't have a budget, you should be able to calculate what the costs of teaching and training are. Identify the number of hours you take to teach a course, and multiply it by your wage per hour figure. For instance, if I make $31,000 per year, my wage per hour would be approximately $15 ($31K divided by 2,080 hours, a typical work year). If I taught 20 two-hour courses each year, the cost of my teaching would be $600 (ignoring benefits, although they should be factored in as well).

But there are more costs involved in teaching and training. Usually they are divided into development and production costs. Development costs include all the time that goes into the analysis, design, and development of a course. It is important to keep track of these in order to get the most accurate view of total costs. It is useful to keep a log of how much time you, and anyone assisting you, spend identifying, collecting, and analyzing data; sketching objectives and storyboarding designs; and actually developing learning modules outlines and instructional materials. Typically,

Fig. 6.3: Sample Budget

Development		
• Design • Development	30 hrs. × $15 = $450 12 hrs. × $9 = $108 Sub Cost:	$558
Production		
• Instruction • Registration • Materials	20 hrs. × $15 = $300 5 hrs. × $7 = $35 800 × $0.25 = $200 Sub Cost:	$535
Staff Time		
• Attendance absence • Substitute	200 hrs. × $11 = $2200 20 hrs. × $6 = $120 Sub Cost:	$2320
Total Cost of Training		**$3413**
• Per employee		$34.14

the time for these activities can range from 30–90 hours to create a course. At $15/hour, a course could cost $450 to $1,350 to create.

Total cost of a course includes all costs associated with one class. Note that development costs are only counted once, even though the course may be offered 20 times in a year. Teaching or training costs are counted each time. Another cost that could be counted is the cost of staff attending training. Take an average of the total wage per hour and multiply it by the number of people taking off work to attend a course. For instance if five people make $9/hour and five make $11/hour, the cost of them attending a two-hour course is $200 ($10/hour average wage × 10 people × 2 hours). That cost is added to the teaching cost of $30 ($15/hour × 2). The development cost would either be added in total to the first class, or averaged over the number of courses altogether.

In addition, there are likely to be production costs. These are usually separated into materials or supplies and personnel. The materials or supplies costs include all of the costs of distributing materials (copying, the paper itself, etc.). These costs are fairly straightforward (for instance, $0.25 per sheet for a handout would include paper, ink, and copier costs). Personnel costs include all the costs of people involved in supporting the course (the people doing the copying or registration, the instructors, etc.). These vary directly according to the wage per hour and amount of time of the people involved. An alternative, or sometimes supplemental, cost can be attributed to purchase of materials that are developed elsewhere, such as "how-to" or "dummy" texts or workbooks.

Often budget considerations are considered implicit, or simply absorbed by the larger general budget of the library. For instance, it may be required that all reference librarians teach 20–40 hours per year, and thus salary is already applied to teaching. You might have a person designated to develop instruction materials, or again make that a part of everybody's responsibilities. You may have a predetermined departmental supply and expense budget that covers all photocopying (but you have to do it yourself). Keep track of the time anyway—you need data to make proposals and at some level within administration budget numbers are needed to support a case, define a benefit, or demonstrate impact.

When developing a teaching or training budget, try to project all of the costs mentioned above—development, production, and implementation. Estimate hours spent in each area, and make a few projections as to the number of courses that might be taught. For instance, show the costs involved in teaching 20 sessions and for teaching 60 sessions. Show where the costs are levied, especially when the work will be spread throughout the organization (i.e., other departments). There may come a time when it is politically advantageous to carve out a specific teaching/training line in the budget, and the more information you have, the quicker you will be able to put all the pieces into place.

SIGN-OFF AND BUY-IN

Rare is the program or initiative that doesn't require some kind of "sign-off" or permission from authority at some level. This may be figurative or literal. Figuratively speaking means that enough general approval has been given to make it okay to proceed. Literal sign-off may require a memo or document that states that the board, director, dean, or department head approves. Sign-off may be general and one-time, or needed before a new phase or large commitment.

As noted above, the best place to start is to get people involved from the beginning. You can sell administration or management by approaching them with questions and concerns. Build rapport and bounce questions or ideas off them. Be positive. Rather than say, "Our system is so hard end users can't navigate it," try to express your concerns positively by saying, "We have an opportunity to increase use of our resources." Build relationships with middle management so that you can ask them to take your ideas forward.

It is possible to get sign-off to go a certain distance, and then have to stop and make the case to proceed. This is a possible negotiation technique if administration or management is leery of backing a complete plan. For instance, it might be possible to get permission to do a survey, but have to make a presentation on results before proposing a next step. Or it might be necessary to create the first of several courses, offer it, and then continue based on the results. Another technique is to offer an abbreviated version of the course to administration and management so they can see firsthand what it will be like.

REVIEW A TEMPLATE FOR BUILDING A TECHNOLOGY TEACHING PROGRAM

Scan the following as checklist for comprehension of this chapter on building a program. Also, you can use this list as a quick checklist when you are building or adding to your own technology teaching.

Can you identify who should be involved in building it?
- Have you identified departments that should be involved?
- Can you identify individuals with whom to liaison?
- Have you touched bases with these individuals?

Have you accounted for factors comprising an "environment review"?
- Are you in line with the library's mission and overall goals?
- Can you articulate general teaching/training needs that match them?
- Have you identified all variations of technology, and the hardware or software problems and needs using them?
- Can you identify specific courses or classes to address them?
- Have you reviewed programs that went on before to identify successes and failures from which you can learn?

How will you prioritize teaching or training needs?

Have you sought as wide a range of input as possible? From staff and end users?

Can you identify primary needs? Secondary? Tertiary?

Can you show linkages between various courses to show how priorities fit together?

Have you emphasized how they meet or impact the library's mission and goals?

What budget requirements have you been able to determine?

Can you identify who will be involved in various aspects? How much time?

What are the calculated costs for development and production? Teaching?

Can you project costs and time for a series of courses to fill a program?

Where and how will you be able (or need) to achieve sign-off and buy-in?

Do you know who is involved?

Can you contact them directly? Use an intermediate to contact them?

Have you made a "sales pitch" to get approval?

7 MAKING THE TECHNOLOGY TEACHING PROGRAM WORK

In this chapter you will:

- **Describe how to use a technology teaching program**
- **Identify ways to market and promote a program**
- **Discuss how a program can be implemented**
- **Define ways to make a program applicable**
- **Identify elements to evaluate a program**
- **Review template for making the technology teaching program work**

INTRODUCTION

Once you've created or built up a program, how do you sustain or manage it and keep it going? You need to apply it to your library's situation and circumstances. Often, what seems good in theory needs to be adapted to work, let alone to work well. To develop participation you need to market and promote the program. Marketing defines what the program is, whom it is for, and what it does for them. Promotion identifies specific actions you can undertake to achieve the goals of your program. Once you have the audience, you need to be able to implement the program successfully and ensure its outcomes apply to participants, as well as to the mission and goals of the library.

UTILIZING A PROGRAM

Developing your program means creating a series of processes, applications, files, and tools to make it into a useful system locally. These include a mechanism for scheduling classes, instructors, participants, and a room with needed equipment. It also includes a system for producing materials, such as promotional fliers, handouts and exercises, evaluations, and feedback, etc.

Creating a schedule for courses and individual classes is no easy matter. You have to have an idea of when is a good time, which you get by asking, checking, and calculating. On the smallest scale, you have decided how long a class should be. The hour-long class seems to be the magic unit, but it isn't always the most effective. Obviously, if you are constrained to one hour you'll have to minimize how much you can cover! Sometimes an hour and a half works well, but two, three, or four hours may be better. Remember, you should always aim to have some kind of break for anything over one hour, whether that's a simple ergonomic exercise break ("Everybody stand up and do shoulder rolls with me"), a quick bathroom break, or a proper 10–15 minute break. If you block out more time just because you're not sure how long things will go, be careful—some people will appreciate getting let out early, but others will feel cheated. Determine the length of the class by balancing how many objectives you need to cover with participant attitudes.

Mornings are usually better for learners, who are alert and eager (as opposed to afternoons when they can be a little lethargic and grumpy). However, mornings are usually the busiest time—you may need to offer some classes early and some later, with the understanding that afternoon sessions may drag a little. Sessions given over the lunch hour are always "iffy" because some people always have the time blocked out for lunch—never make mandatory training required at noon! Mondays and Fridays tend to be less attended than other days of the week for obvious reasons (people are just getting started with or winding down their week). Obviously, you have to look ahead to ensure training doesn't interfere with already planned events, meetings, holidays, etc. But you can't please all of the people all of the time. Do a little groundwork, survey some users as to their preferences, and try to accommodate your and their time as best you can.

Whatever schedule you come up with, it is important that you project it far enough ahead that people have time to deal with their own schedules to assure their attendance. At a minimum, one full month is needed. Thus, if you're planning training for March, you should have the dates scheduled and announced sometime in January. More often, you'll work your schedule based quarterly on seasons (i.e., spring and summer courses) or half year semesters (i.e., fall and spring courses). Thus you'll need to plan four to seven months ahead. The more time, the better to advertise and promote the classes.

Fig. 7.1: Sample Course Schedule

Courses in June 2002	Date/Time/Class size	Date/Time/Class size
Mousing Skills	6/3/02 10 am (10)	6/10/02 1 pm (10)
Basic Windows, Sect. 1	6/5/02 10 am (15)	6/12/02 2 pm (15)
Basic Windows, Sect. 2	6/19/02 2 pm (15)	6/21/02 10 am (15)
Searching the Catalog	6/13/02 11 am (15)	6/18/02 1 pm (15)
Internet Search Basics	6/7/02 1 pm (10)	6/17/02 11 am (20)
Advanced Internet Searching, Sect. 1	6/24/02 10 (15)	6/26/02 2 pm (10)
Advanced Internet Searching, Sect. 2	6/27/02 2 pm (20)	6/28/02 10 am (20)

Fig. 7.2: Sample Course Promotion

Are you ready for summer! In addition to school vacation and other time off, have you thought about taking some leisure time to hone a few new computer skills? Your Library is once again the relaxation spot to learn some new tips and techniques. Our June schedule includes courses in Internet Searching, as well as refresher courses in mousing and windowing skills. Register now! Check our online page at: http://www.lib.library.org/courses. Or phone us at 123-456-7890. Or stop by and talk to one of our staff—they'll be happy to discuss the courses and sign you up!

A big issue for some people is whether to require registration of participants in the first place. Keep in mind that registration is a matter of both philosophy and logistics. Philosophically, some people like to leave attendance wide open so that anyone could drop in at anytime. However, registration solves a lot of problems—you know how many people to expect and how many handouts to produce; the attendees have made a commitment to be there; you have a list of names to work with; and you have had an opportunity to screen them regarding prerequisites and objectives to be covered. If you wanted to accommodate drop-ins, you can, just don't advertise it. Logistically, just as with scheduling, someone has to be available to do it. You may need to require registration at certain times when you or someone else will be there to answer the phone. Using e-mail or Web form registration is a viable alternative to the phone, but still takes time to type or transfer names to a list. Once you have captured registration information, you are already keeping statistics. So think about what kind of "demographic" data you want to include as part of registration (age group, geographic distribution, etc.).

Depending on how many instructors you use, scheduling them may be simple or complicated. This presumes that you have iden-

Fig. 7.3: Sample Registration Form

Registration for June 2002 Summer Computer Courses

Name (last, first)	Marge Brauer
Phone number	123-345-4567
Address	5291 Elmwood Ave. Shelby Twp., IN 47907
email	margieb39@hotmail.com
Course(s) desired (in order of pref)	
Skill level?	Novice Intermed. Adv. Expert (please check one)
Your first time?	Yes No

tified instructors who are willing and able to be a part of the training—sometimes you have to make the hard decision to forego the use of instructors who are available but not effective. Instructors should know far enough ahead of time to prepare for classes. If they do not actually develop the instruction they will need time to practice. And whenever possible they should show up 15–30 minutes to test out the room, ensure materials are distributed, meet attendees, etc. As with reference or any other service, you may need back-up in the event someone has an emergency. Use of a centralized or shared calendar to keep track of instructor scheduling is helpful, but be sure to contact instructors ahead of time as a reminder and to check that everything is all right.

Another part of scheduling includes booking the location (area, room, building, etc.) and equipment. Of course, this presumes you have already negotiated permission to use these facilities in the first place. Obviously, if you are going to take over the workstations dedicated to online catalog use in the reference area you'd better check with someone ahead of time! It's so nice to have a dedicated area or classroom with workstations and a computer projector, but if you don't have them, you have to make do. If you have to use public workstations, be sure to negotiate with whoever is in charge of the area and advertise both on the day of the training and during the week ("This area will be reserved for training 10–11 a.m. every Monday—please contact the reference desk for searches in the online catalog at that time"). If you are using your own facilities you'll probably have less work to do than if you are using someone else's. Be sure to build up a relationship with whoever is in charge of the equipment because you will be working with them a lot when the training becomes successful and you need to add more sessions.

A final, but nonetheless important, part of scheduling is the production of handouts, exercises, evaluations, feedback forms, etc. If you run a one-person shop, try not to fall into the trap of producing copies just before a class—as with anything else, you need to allow yourself time in the event of emergencies, such as a broken photocopier. Whenever possible, try to farm out the production of copies to a department (administration) or unit (printing services) that deals with it on an ongoing basis. Aim to have copies ready 24–48 hours ahead of time, and produce a couple of extras in the event of drop-ins to the class.

MARKETING AND PROMOTION

Marketing your program means defining your program: detailing what the program does, who it aims to help, and how it can help them. You need to look at where and how the program fits into the local training market or what it competes with, and then determine how best to "sell" it. Do a market analysis before jumping into design and development. Promoting the program requires creative advertising to get the message out to potential participants. Promotion may be handled differently when working with internal as opposed to external audiences.

As noted, marketing means you first analyze your product in terms of how it can be described and what people currently think about it. For the most part it should be easy to define your program if you have already designed and developed it. Your course titles and objectives provide pretty good descriptions of what you offer. Use these to create clear, concise abstracts to describe your offerings. Formulate them into a course listing or course catalog to which you and others can refer. Identify what you think are the strengths of your courses and highlight them.

But before you offer your courses, you should look around to identify programs or other things with which it competes. Know your market! Are there similar courses offered that participants might be able to take? What distinguishes your classes from others (time, subject, cost, etc.)? Compare your listings to theirs to identify strengths or weaknesses. In addition, look to see what else competes with your courses—books, online tutorials, video or television programs—and try to identify advantages for using your services over the competition. Gather all this information to help you make decisions regarding promotion of your training.

Don't believe in that story line that "If you build it they will come"...You need to actively expend part of your time in this

Fig. 7.4: Internal E-Mail Promotion

Date: Fri, 14 Jun 2002 08:13:39 -0500
From: "SCOTT BRANDT" <techman@purdue.edu>
To: "All Libraries" <all-lib@purdue.edu>
Subject: Summer Training Courses, July 10, 2002

Hello,

I wanted to announce that the summer catalog of courses is up on the Staff Development and Training Website <http://www.lib.purdue.edu/staffdev>. This summer you'll find a lot of new courses in both training and professional development. In particular, we're offering a new Project Management Course ("Planning Your Project"), new Library Resources classes (Searching... For Statistics, News, and Current Events), and one that several people have asked about, "Converting Handouts to the Web." Other courses include "Personal Strategic Planning" and "Applying Tactics of Innovation."
Be sure to access the course registration page and submit registrations no later than Friday June 28. As always, objectives for the courses will be linked to the Staff Dev site so that you can preview the courses and assess your needs.

Hope to see you in class!

Scott

process sending out notices and reminders, creating commercials and advertisements, doing demos of what's coming up, and convincing people of why they want or need to come. You need to determine how proactive you can be. And you need to decide if you can do it, or if you must pass this on to another unit or department. Whether you like it or not, you need to sell your services (or have someone to do it for you).

One of the biggest places to advertise is within the library—on the Web site or anyplacc where potential participants gather (circulation/reference desk for patrons, newsletter for staff). Use this free opportunity to describe and talk up the courses offered. As with any promotion, consider using a variety of formats and approaches, such as bookmarks, fliers, Web pages, e-mail, newsletters, etc. Small calendars with pertinent dates would be handy for potential participants. Descriptions of outcomes, or what can be achieved with the training, are also useful.

Obviously, you have to be a little careful when advertising outside the library. Often, when you provide descriptions to others they may have to edit them. Try to anticipate this by providing both short and long descriptions, and perhaps serious as well as funny ones. Where appropriate, consider using avenues that will find potential participants outside the library, for instance as a public service announcement on television or in conjunction with other community events (bazaar, flea market, festival, chamber of commerce activity, grocery store bulletinboard, etc.). For internal services to staff, try to be just as creative. Send e-mails that resemble commercials, public service announcements, etc.

As part of your advertising, consider using the 3 Ds—Describe, Demo, and Differentiate. Make your descriptions interesting but factual. Describe what is offered, but also what can be achieved as an outcome. For instance, an e-mail course could be described in terms of its objectives, but will be more enticing if you note that pictures can be sent as attachments (as is appropriate). Demonstrate outcomes wherever possible. Remember the old adage that a sale is more successful when customers can touch the merchandise. In lieu of physically touching something, let them see it in action—use a personal testimony, or in the case mentioned above, show the picture of a grandchild that can be sent in an e-mail. Differentiate, or compare and contrast, between what happens if someone takes the courses versus if someone doesn't. Use scenarios to describe successes or failures, highlight costs and benefits, or detail opportunities lost or gained.

IMPLEMENTING THE PROGRAM

Implementing the program means rolling up your sleeves and getting to work. As a result of your marketing and promotion you will have customers whose needs you must meet. You must turn your focus to the classroom, which becomes your workroom, and on the interaction between the main elements—effective trainers, pertinent training, and motivated attendees. First, determine how the physical elements (a room, equipment, materials) necessary to accomplish the training will come together. Then decide how trainers will be prepared to do the training. Last, look at how the program can help ensure learners are motivated and eager to learn.

A well-prepared training environment is a direct consequence of a well-executed program. The room, or environs, used for training will have been booked at the time the schedule was developed. Someone should check the room prior to scheduling to verify that it is a good space for teaching or training. Ideally it will be an area closed off from distraction. Preferably the room will allow good spacing—learners will not be crowded and the instructor will have room to move about freely while talking and doing demonstrations. Hopefully, computer monitors will not block eye contact between the instructor and learners, as this is a critical method for checking in on learners.

Even though the room has been previously looked at, instructors should call the day before to make sure it is available and show up 10–30 minutes ahead of the training to check it for each session. This really serves three purposes: First, it allows time to check for major changes or problems. If the room can't be used, there may be time to contact participants to cancel (depending on the setting). Having equipment turned on as attendees show up allows more time to be spent on introductions and preliminary instructions. Second, it ensures that the instructor is there to meet students. There is nothing worse than instructors showing up late—it makes them look bad, and usually creates a "hurried" atmosphere. Third, it requires that the instructor be prepared well ahead of time. Instructors should block out time to review notes or practice examples two or three days ahead, not just before class. Likewise, as mentioned earlier, handouts and other materials should also be prepared well ahead of time.

As noted in an earlier section on trainer traits, in addition to good delivery, instructors should be well organized. Even once the design and development of the instruction has been done, trainers need to check and review modules as part of implementation. Objectives must be looked at to determine if they are still appropriate and fit the needs of this particular group of participants. Examples must be reviewed to ensure they are pertinent and still work. If Web pages or Web sites are included, it is imperative that they are checked to ensure they can still be shown. All of this must be done prior to the class since changes need to be reflected in handouts, exercises, and other materials. Good trainers develop a system where they review material four or five days prior to a class, get materials printed up a day or two before, and still show up early on the day.

Typically, a session proceeds smoothly with a beginning, middle, and end. The beginning involves "housekeeping work" as well as providing any needed introductions. If used, a registration list records participant attendance and helps identify them by name. Handouts will be distributed. Instructions regarding equipment are given. The end is a wrap-up that may also include some activities not directly related to the training. There will likely be an evaluation form to fill out. There may be additional materials to distribute. Appointments may be made for follow-up or additional assistance.

The middle part of the session is likely to take up the bulk of the time. The objectives will be covered and outcomes sought. It is best to start the training by describing what the session will cover. It is important to link this to any previous learning or knowledge of the participants. Then begin covering the individual objectives with the strategies you have chosen. Be sure to include opportunities for each objective to be practiced or somehow reinforced for the learner (perhaps by mini-quizzes or reviews). End the main training by reviewing the overall learning and having participants demonstrate the outcomes.

Remember that learners are best motivated when the learning is attached to what they know and pertains to them. One place to gain insight into the composition of the group is by chatting with them before class. When you make introductions and connections you discover some of the background of participants that you can use in the training. For instance, if someone mentions that they come from a rural community, you might be able to use examples relating to agriculture. Or if someone notes that her company builds warehouses you might be able to use architectural examples. This also helps personalize the session. You might say, "I was talking to Bob earlier and he said he hoped we'll be covering search engines—and I noted that we would be looking at two in particular."

Some people have to work with a script, while others feel free "just winging it." Don't be so rigid that you don't allow for unforeseen interaction, but don't be so laid back that your objectives aren't accomplished. If you must use an outline, try to make it general—don't read from a script. Outlines are actually good for both rigid and laid-back instructors. Hopefully when it comes to doing the training, you'll find a good mix of structure and flexibility.

Also, remember the old adage, "Keep it simple, stupid." Don't complicate the session with extra examples, stories, or exercises to impress the participants. Try to stick with simple, concrete examples. Keep in mind that Einstein didn't use formulas to explain relativity to lay persons, he used an analogy of two people passing each other on a train. Don't get sidetracked with multiple examples or personal stories. Don't force participants to do several exercises "to ensure they get it."

MAKING THE PROGRAM APPLICABLE

An important point about any program is that the instruction or training program doesn't exist in a vacuum or only for itself—to be successful it must be applicable to the learners' needs. To meet their needs, it may be necessary sometimes to provide training on demand. For staff, it often requires tying the training program to promotion and performance. For customers, patrons, or students it requires tying the program directly to their perspective.

There is a sticky problem that instructors often face: should training be scheduled and administered within a well-developed structure, or should it be offered whenever needed? It makes sense, once you create a teaching or training program, to use it as the basis for scheduling instruction. By breaking down the system into phases (analysis, design, development, etc.) and scheduling all the work with a long lead time, you can line up instructors, advertise courses, and register participants with no pressure of time constraints. However, there will be times when you have to accelerate development in order to accommodate learners, for instance, when new software suddenly shows up! Keep in touch with developments so that you have as much lead time as possible, but be flexible—perhaps once you have the training developed you can add additional sessions or slightly alter materials at the last minute to meet "just-in-time" demands.

Staff must see the training program as a tool to improve their situation. They must feel the program is "blessed" by management, who must support the program by actively encouraging time off. It must be indicated that workloads can be altered to provide sufficient time to take the classes. When done successfully, individuals will feel empowered to act on their own. Otherwise the program will not become institutionally recognized and accepted. It helps if management will do more than give "lip service" to such practices—for instance, by mandating that 30 hours per year be spent in staff development or continuing education courses.

For customers, patrons, and students, a program has to be seen as both an integral part of regular library service and their own lives. First, they must see it as a natural outgrowth of using the library, otherwise they may not take to it. For instance, they may not believe that librarians are competent to teach "how to search the Internet," but they will increase their faith once it is demonstrated that searching catalogs and indexes is similar to searching on the Internet. This is not just a matter of promotion, but of actually convincing them that the program is a solid service. Second, they must see the program as an extension of their own curiosity and desire to learn. That is why it is important to get their input at all phases—beginning, during, and after. Learn from the reactions of participants. Ask for their opinion on various parts of the program to get a view from their perspective.

Fig. 7.5: Formal Program Evaluation

Please check the response that best matches your opinion	Strongly agree	Agree	Neutral	Disagree	Strongly disagree
Overall, the program meets my needs					
I am able to apply skills learned					
• During training course					
• Shortly after attending					
• For a long time afterward					
More searching classes are needed					
• Searching Library resources					
• Searching specialized indexes					
• Other Internet search engines					
Timing for the schedule is good					
• There is variation through the year					
• There is variation in day of week					
• There is variation in time of day					

EVALUATING THE OVERALL PROGRAM

Just as with individual training courses, the program as a whole must also be evaluated. Formative evaluations take place while putting a program together, summative refers to evaluation after the completion of the program.

It is usually hard to get useful formative evaluation feedback from participants, unless it is over a long period of time. For one thing, they are probably unaware that the program is in a formative stage unless you tell them. For another, the feedback is likely to be about the individual courses they have taken, rather than the program as a whole. Colleagues and co-workers are more likely to give you formative input. They might be able to comment on the ways to streamline copying and distributing materials, or on processes related to registering participants or checking on equipment. For instance, a colleague who has extensive experience working with a database may show you an easy way to capture registration information from a Web form directly into a database.

Fig. 7.6: Informal Program Evaluation

Date: Fri, 14 Jun 2002 08:13:39 -0500
From: "SCOTT BRANDT" <techman@purdue.edu>
To: "Jennifer" <jshark@purdue.edu>
Subject: Summer Training Courses, July 10, 2002

Hello Jennifer,

I wanted to ask if you would respond to a couple of general questions regarding the training program in the Libraries. I know you attend several courses and send many of your staff as well. Could you identify a few areas of possible improvement that we might try to address for next time? Any skills we're overlooking in your unit? Any new software on the horizon that could use training? Any courses in particular we should repeat? Any anecdotes you could share about successes or failures in the classroom or transferring skills to the workplace?

Thanks in advance,
Scott

Summative information may be a little easier to elicit from participants after they have been through a few courses. They would then be better prepared to comment on trends or patterns of problems in registering, checking prerequisites, using handouts, etc. It might be useful to seek this information at the end of each cycle (e.g., yearly or by semester). You might ask specific questions, "Was the time between courses: A. Too short; B. Too long; C. About right; D. Didn't matter?" or open-ended ones, like "How easy was the registration process?"

Just as with the course evaluation, you can evaluate the overall program in several ways. You can try to formally measure feedback or do informal follow-up surveys. You can elicit information directly or indirectly. Formal measurement is going to require quantitative analysis. Remember, that simply means collecting data ("Choose the response that best fits your attitude: 1. Very strongly; 2. Strongly; 3. Neutral; 4. Weakly; 5. Very weakly?") and synthesizing it ("30 out of 40 participants responded they were strongly or very strongly pleased"). Informal surveys don't require filling out forms, but anecdotes gathered from them should be recorded. Direct contact with participants will get you subjective feedback based on personal interaction. Indirect contact with parents, teachers, supervisors, etc. may give you additional objective second-hand information.

Another way to evaluate is to analyze program-related data. The most obvious number to look at is attendance. After initial offerings, full classes are a good indication of success, often due to word of mouth. A closer look at registration and attendance can identify other trends, such as repeats or people who work their way from introductory to advanced courses. Similar data also helps identify popular courses for which you might add sessions and those that should be canceled. Indirectly related data to track might include increases in the number of indexes or other resources used.

TEMPLATE FOR MAKING THE TECHNOLOGY TEACHING PROGRAM WORK

Scan the following as checklist for comprehension of this chapter on making the program work. Also, you can use this list as a quick checklist when you are fine-tuning your own technology teaching program.

How do you prepare for the technology teaching program?
- Have you lined up instructors?
- Do you have access to rooms and equipment?
- Have you looked at variations in scheduling times? Have you checked for conflicting dates or events?
- Are you using registration? Online, over the phone, or in person?
- Have you prepared handouts and other materials far enough ahead of time to allow for proofreading and reproduction?

What will you do to market and promote the program?
- Have you analyzed your "product" to better describe it in terms that users will understand and appreciate?
- What other programs or events compete with yours?
- How many kinds of marketing are you utilizing? Ads, newsletter descriptions, preview demos, e-mail notices?
- Have you defined a description, prepared a demonstration, and identified how to differentiate your program from other competing events?

How will the program be implemented?
- Have you reviewed and booked necessary equipment and rooms?

Will instructors be able to get into the room 15–30 minutes prior to setup? Can they bring up software applications and Web pages? Have they double-checked the examples they plan to use? Will they greet attendees to get to know them?
Is there a plan for conducting the course? Do instructors review "housekeeping" details, review objectives, allow practice, and summarize?
Do instructors have an outline or script to follow?
Have you tried to remove complexity in demonstrations and exercises?

Where will you make the program applicable?
Have you determined how the participants will use the training?
Will it be offered on a strict schedule or when trainers want it?
Has management endorsed the program?
Are participants empowered to sign up? Do they see courses as natural services of the library? Is their curiosity and growth encouraged?

Which elements will you use to evaluate the program?
Are you getting program information from participants, coworkers, or the program itself?
How will you seek formative data while the program is developed, or summative data after it is completed?
What kind of course-related data will you seek?
How will you use the evaluation data you compile?

3 EXPLORING SUCCESSFUL PROGRAMS

In this chapter you will:

- **Review aspects and successes of various library teaching/training programs**

INTRODUCTION

To give you a general idea of the kinds of programs offered at a variety of libraries (large and small public, corporate, large and small university), I have included this section. In 1993 there weren't a lot of technology training programs across the country, but there are more now and the number is constantly growing. Entries are presented here as quoted from the individuals.

In addition to getting an idea of the scope and coverage of these libraries, a sample of some of their greatest successes is included to inspire you. Note that several of the proudest accomplishments include growth and taking on new responsibilities. I hope you will be encouraged to do likewise. Some of these achievements may sound simply like features, but also reflect the successes of incorporating them systematically into a program.

SCIENCE, INDUSTRY, AND BUSINESS LIBRARY, NEW YORK PUBLIC LIBRARY

The New York Public Library
Science, Industry, and Business Library
188 Madison Avenue
New York, NY 10016
www.nypl.org/research/sibl/index.html

Contact person for the instruction program:
Janet Bogenschultz
Training Coordinator
jbogenschultz@nypl.org

KIND OF TRAINING/TEACHING OFFERED

- By offering 15–20 classes weekly to the public, at no charge, SIBL promotes the use of diverse electronic and print resources in its Electronic Training Center. See biweekly schedule at www.nypl.org/research/sibl/training/.
- Currently, SIBL has a curriculum of twenty classes in the public training program. We had two new classes in September, and will have two new classes in October/November.
- Library Resources: There are five class types. Four class types are focused on library resources: Library Skills (e.g., catalog searching), Science, Business, and Government Information. Library resources classes include 15 classes in our curriculum.
- Four Web classes in our curriculum.

MOST POPULAR CLASSES

While our four Internet classes continue to have broad appeal, and are popular with the general public, NYPL's Branch Libraries also offer instruction in this area to NYC's general public. As a result, the SIBL instruction program devotes more time and energy to 15 classes focused on SIBL's resources in its collecting areas of science and technology, industry, business, and government information. Titles of the most popular include:

- Stocks & Mutual Funds: Investment Resources @ SIBL
- Introduction To Patents
- Market Research Information Sources
- Directories: Using Them To Find Companies and People
- Articles: How To Find Them Electronically

The titles of our popular Web classes listed in the order of their popularity:

- Web Workshop for Beginners
- Web Workshop 2: Search Engines
- Job Searching on the Web
- Web Workshop 3: Evaluating Resources

TO WHOM TRAINING IS OFFERED

- Library staff
- The public
- Students from both high school and NYC area colleges and universities
- NYC businesses (customized per request) and small business owners attending public classes

- Library donors
- Educational organizations
- Community groups
- Entrepreneur service providers
- Requests for customized group instruction are reviewed on a case-by-case basis, and may be fee-based, depending on the group

HOW OFTEN COURSES ARE OFFERED

Our public training program consists of daily classes, six days a week:

- Monday–Friday, three to four classes per day
- On Saturdays we usually offer one class, except for during the summer months
- Frequency of a particular class may vary
- Most individual classes in the curriculum are offered weekly, with some offered once every two weeks

SESSION SIZES

- We teach small to large classes, with ten being the average class size since May 1996 (opening date).
- The capacity for a typical public training class is 17. The first 13 attendees each have their own workstations, with room for four "waiting list" attendees observing at the back of the classroom. If registrants are no-shows, attendees on the waiting list may move onto the workstations.

TYPE OF FACILITIES USED

Dedicated classroom. We have four hands-on training classrooms, each with 13 workstations, in our Electronic Training Center. Our public workstations are not used for training.

MOST SUCCESSFUL TECHNIQUES

- Hands-on exploration of complex resources with the instructor, highlighting features, executing searches, demonstrating refining/revising searches and incorporating ideas/questions, followed by hands-on exercises, often based on "real" reference inquiries or class participants' interests
- Accommodating the variety of student computer skills and interests with post-lecture practice time where the instructor provides one-on-one assistance at each workstation. Web 3: Evaluating Resources has group exercises on the evaluation of Web sites based on the established criteria

introduced in the class and encourages participation, and critical thinking.

- Unique exercises: The following worked well in a team-taught lecture/demo class called "Point & Click to Success: Finding Small Business Web Resources," in which having individual exercises is problematic. The class was created for offsite instruction.
  1) Each attendee was assigned a fictional business with brief background, and a current business challenge
  2) At the end of each section of the class, the instructors asked if any attendees had business challenges that could be solved by research using the resources just discussed
  3) The "business challenge" approach engaged participants in discussion and the hypothetical application of resources, without the aid of a computer
  4) This exercise, based on 30 different small-business challenges, was further modified into an exercise handout that attendees could use to explore small-business Web resources after class on a computer at home or in their offices
  5) As this was offsite instruction in a rural area, and Internet connectivity had the potential of being problematic, the Power Point slides contained numerous screen shots of small-business Web resources. By anticipating this problem, the two sessions without connectivity were extremely successful.

PROUDEST ACHIEVEMENT(S)

- Prior to the opening of SIBL in May 1996 many of the Information Services (IS) librarians had little or no experience in formal classroom training or curriculum development. Now, five years later, all IS librarians teach and help to develop new and revised existing courses. The result is a roster of roughly 20 classes (as new ones develop, some are retired) which to date have reached more than 50,000 "walk-ins" who register on their own.
- Staff continues to be enthusiastic in their teaching, and feel a group ownership in the instruction program due to their input and involvement.
- Every class session receives a formal written evaluation. These are most often positive. Constructive criticism and suggestions are taken seriously. Respondents' expressed research needs form the basis of new classes.
- Requests for customized training continue to increase. This

year, 60 groups arranged for instruction tailored to their specialized needs. These included sessions on: advanced marketing research for university courses, advanced Internet searching for a B2B advertising agency whose staff relies on the Internet for identifying potential customers, researching product packaging design for college students, science resources for NYC science high school students entering the Intel Westinghouse Competition, and business resources for entrepreneurs that several small business service providers arranged for their clients.

- An index to the librarians' enthusiasm for teaching is that they now develop new classes to complement SIBL's exhibits. For example, librarians developed a special class on astronomy resources tailored both for the general public and for school groups to run during the showing of "Heaven's Above: Art & Actuality," which contrasts the nineteenth-century art and science of Etienne Leopold Trouvelot with contemporary photographic images from NASA at www.nypl.org/research/sibl/trouvelot/.

UNIVERSITY OF TEXAS AT AUSTIN GENERAL LIBRARIES

General Libraries
University of Texas at Austin
Austin, TX 78713
www.lib.utexas.edu

Contact person for the instruction program:
Elizabeth Dupuis
Head Librarian, Digital Information Literacy Office
beth@mail.utexas.edu

KIND OF TRAINING/TEACHING OFFERED

- Primarily library resources and Internet
- Technology (Word, Windows) is taught by the campus computing department, ITS

MOST POPULAR CLASSES

- Course-integrated classes are skyrocketing by requests from all departments, especially Freshman Seminars
- Also offer prescheduled classes to which anyone is wel-

come for fall and spring, especially UT Library Online (our Web site), e-books, and EndNote (bibliographic software)
- Faculty workshops, especially plagiarism and online library services

TO WHOM TRAINING IS OFFERED

- Staff (including faculty), the public, and students
- We do not conduct regular sessions specifically for businesses and we do not charge for any of our classes but people from businesses are welcome to come as "the public"

HOW OFTEN COURSES ARE OFFERED

- Course-integrated classes are taught, often more than five per day
- Pre-scheduled workshops are offered once per day every day through the first full month of school during fall and spring semesters
- Faculty workshops happen the week before each semester begins
- Staff training occurs sporadically as issues arise

SESSION SIZES

- Course-integrated classes are large (20+)
- Pre-scheduled workshops range from small (<10) to medium (10-20)
- Faculty workshops are often medium (10-20) (occasionally <10)
- Staff training is usually medium (10-20)

TYPE OF FACILITIES USED

- All classes occur in dedicated classroom with hands-on equipment, though some occur in another room which is lecture/demo style
- Much other training happens one-on-one on public workstations but that is classified as reference, not instruction

MOST SUCCESSFUL TECHNIQUES

- We use active learning activities in most course-integrated classes, though more often in undergraduate than graduate-level classes

PROUDEST ACHIEVEMENT(S)

- Last year we reinvented the pre-scheduled workshops from 150 offered each semester to about 30. We stopped teaching the HTML classes once campus seemed to have gained these skills.

- We focus more of our instruction energies on course-integrated classes. We also partner with the ITS department whose role is to teach applications and platforms—Windows, Word, Photoshop, Dreamweaver, SPSS, etc.
- Additionally we designed an online tutorial to ensure students in the freshman level classes had a basic introduction to research skills and library resources so our instruction could focus less on basic issues about technology or libraries and more on searching, evaluating, and relating to the specific course they are taking.

ITHACA COLLEGE LIBRARY

Ithaca College Library
Ithaca College
Ithaca, NY 14850
www.ithaca.edu/library/

Contact person for the instructional program:
John R. Henderson
Reference Librarian
Ithaca College Library
jhenderson@ithaca.edu

KIND OF TRAINING/TEACHING OFFERED

- Library resources and Internet

MOST POPULAR CLASSES

- Library Resources and Methods of Research. A one-credit, seven-week course. The course is designed to enable students to better use basic library materials, online resources and reference tools; to formulate and clearly define a research topic, plan a search strategy; and to critically evaluate research materials.
- Workshops offered as part of the Faculty Development Colloquia Series. Usually once or twice a semester we offer a workshop featuring some aspect of the library. Past topics have included "Periodical databases" and "Is the Web a Good Research Tool?"

TO WHOM TRAINING IS OFFERED

- Primarily students
- Secondarily faculty and staff

HOW OFTEN COURSES ARE OFFERED

- The course is offered twice weekly for seven weeks, three out of four blocks of the school year
- Students also informally daily
- Faculty workshops are offered once or twice a semester
- Other training is offered occasionally

SESSION SIZES

- Ranges from small (<10) to medium (10-20) to large (20+)

TYPE OF FACILITIES USED

Both dedicated classroom and public workstations

MOST SUCCESSFUL TECHNIQUES

As part of the one-credit course and other workshops and session, the ICYouSee: T is for Thinking: The ICYouSee Guide to Critical Thinking About What You See on the Web (www.ithaca.edu/library/Training/hott.html) has been used. It includes a pop quiz that compares a legitimate and bogus page related to AIDS statistics, plus a homework assignment comparing either Web sites related to the Sixties or the Mayan Calendar.

PROUDEST ACHIEVEMENT(S)

I have been very pleased that ICYouSeeT is for Thinking, which I designed for teaching my own students, has been incorporated by teachers, librarians, and other trainers all over the world (if Australia, Canada, and Israel and several states in the U.S. can be defined as all over the world) and for so many different age groups, from middle school students and young military academy cadets to college and university students (including, I think, some grad students).

BUSINESS RESEARCH CENTER, ANDERSEN (ATLANTA)

Business Research Center
Andersen (formerly Arthur Andersen)
Atlanta, GA 30303

Contact person for the instruction program:
Susan Klopper, Director
(SLA Management Leadership Award 1998)
susan.m.klopper@us.andersen.com

KIND OF TRAINING/TEACHING OFFERED

- Andersen's Intranet
- Internet, various aspects of navigating/tips and tricks for utilizing effectively
- Information Resources—external research databases subscribed to for different practices (e.g., tax, audit, consulting)

MOST POPULAR CLASSES

- Hard to say since many classes are scheduled as part of required orientation or are critical to the practice; knowing how to effectively use the research tools that define the research they do for client engagements
- People probably "enjoy" the classes on the free Web best because they generally learn helpful tips/tricks and recommended URLs that always make their lives and doing their jobs easier

TO WHOM TRAINING IS OFFERED

- Professionals in the practice (our internal customers)

HOW OFTEN COURSES ARE OFFERED

- Approximately every two weeks

SESSION SIZES

- Most classes are 10–15
- Occasionally 20–50

TYPE OF FACILITIES USED

Dedicated classroom

MOST SUCCESSFUL TECHNIQUES

- For the training that focuses on teaching them how to use information research tools specific to the technical research needs of their respective practice (tax databases for example), the approach that has proven to be most effective is to team up with an experienced senior person and have them develop case studies and examples which are woven into the course presentation
- I also like to have seniors present for the class since they can field questions and provide "real life" situations which highlight both the relevancy of the database to their work and the importance of focusing on more than the "mechanics" and striving to become a creative and fluid research thinker/searcher.
- If the senior is not able to be present (which is most of the time), I have been working with the practice long enough that I have a good enough understanding of who is working on what so I can work the group to draw out relevant experiences.
- Essentially, it is helping the practice to understand the importance of differentiating between a mediocre search and one that yields substantive leads and results, the latter of which will better meet the expectations of their partners and our clients. Since I am able to demonstrate through my training that I am more adept at using these interfaces than they are, I reinforce my value to them as well.

PROUDEST ACHIEVEMENT(S)

- Becoming recognized as the information guru for our market circle and regularly asked to develop training for various groups and applications

ARLINGTON HEIGHTS MEMORIAL LIBRARY (ILLINOIS)

Arlington Heights Memorial Library
500 N. Dunton
Arlington Heights, IL 60004
www.ahml.lib.il.us

Contact person for the instruction program:
Bill Pardue
Electronic Resources Specialist
bpardue@ahml.lib.il.us

KIND OF TRAINING/TEACHING OFFERED

- Library resources (catalog classes for the public. "Online Resources Group" does review of databases for staff)
- Internet (hands-on "Internet Skills" and Internet demonstrations for the public, also specialized hands-on sessions on online genealogy ("Online Resources Group" covers Internet tips and trends for staff)
- Technology ("Basic Mouse Skills" class for public. Staff training includes regular Windows, Word, Excel training, etc.)

MOST POPULAR CLASSES

(Most popular public classes)

- Beginning Mouse Skills—slow-paced class; users learn the necessary computer skills needed to use the Internet
- Basic Internet Skills—moderate-paced class; learn to use a Web browser to enter addresses, maneuver through Web pages and fill out forms and find information online
- Advanced Internet Skills—faster-paced class; use several popular search engines to find information online, use specialized tools (e.g., library databases, online phone directories)

(Most popular staff training)

- Online Resources Group (ORG)—monthly lunchtime sessions for staff in our Training Center to cover Internet topics, overviews of our databases, and online resources, hints, tips, tricks, etc.
- Staff also frequently attend training for Microsoft Windows, MS Office products, Groupwise E-mail, etc., as needed. This training is done by our staff trainer.

TO WHOM TRAINING IS OFFERED

- Staff, the public, and students

HOW OFTEN COURSES ARE OFFERED

- Not quite daily, but more than weekly

SESSION SIZES

- Hands-on training in small classes (up to eight) for both staff and public
- Demonstration training can have 20–100 people

TYPE OF FACILITIES USED

Dedicated classroom, but some demo only

MOST SUCCESSFUL TECHNIQUES

- Keeping curriculum short and simple, limiting the range of material covered, providing ample practice time at the end of hands-on classes (a 1.5 hours class should have 30 minutes allocated for practice time)
- On the planning side, it's been most useful to have staff submit comments after each session they teach. Instructor comments are by far the most useful input we have for revising the curriculum. Staff submit their comments via an online form. They are later compiled in a report that the curriculum design group can review.

PROUDEST ACHIEVEMENT(S)

- Working collaboratively with a group of staff to identify our instructional objectives and to create well-designed class guides and handout materials. The goal was to provide a set of materials that would almost allow an instructor who is unfamiliar with a class to come in and teach it "cold." Of course, we do review the material with the instructors first, so that scenario should never happen! It's detailed enough to serve as a "script" for those who prefer to work with one, but loose enough to allow other instructors to work through "bullet points" with a lot of their own style.

ST. JOSEPH COUNTY PUBLIC LIBRARY (INDIANA)

St. Joseph County Public Library
304 South Main Street
South Bend, IN 46601
www.sjcpl.lib.in.us

Contact person for the instruction program:
Michael Stephens, Head
Networked Resources Development & Training
m.stephens@gomail.sjcpl.lib.in.us

KIND OF TRAINING/TEACHING OFFERED

- Library resources, Internet and Technology, except we do not teach applications under the technology category, we teach classes like Designing a Web Page and Exploring Digital Cameras

MOST POPULAR CLASSES

- Hands On Basic and advanced are super popular. These are taught all over the system (we published our fall schedule of classes in late August and filled many classes through October within days)
- Other popular classes: Internet and Computer Basics for Seniors (SUPER popular!), Intro to Web Design, Exploring Digital Cameras & WWW Photo Sites

TO WHOM TRAINING IS OFFERED

- Staff and the public

HOW OFTEN COURSES ARE OFFERED

- Staff: We offer a monthly schedule of class meetings for various library agencies. Looks something like this:
  - Adult Reference: Two one-hour sessions per month (second and fourth weeks)
  - Main Service Heads: One one-hour session per month (first Thursday of every month)
  - Branch Heads: One two-hour session per month (first Tuesday of every month)
  - Branch Assistants: One 90-minute session per month (second and fourth Wednesdays—held twice to accommodate all assistants!)
  - Magazines/Newspapers & Fiction: One 90-minute session per month
  - Special Services: One 90-minute meeting per month
  - Children's Services: One 60-minute session each month (first Wednesday of month)
  - Audio Visual Services: One 60-minute session each month (third Monday of month)
- Also schedule special sessions for new staff and special topics as requested

SESSION SIZES

- Mostly small (<10) to medium(10–20)....in 1996, when we started an Internet demo class, we would have 60+...once even 120+!

TYPE OF FACILITIES USED

- Staff Training: we have a ten-computer training lab right now. A new one is in the works for 2002.
- For public classes we use banks of Macs located in the public area or a traveling set of wireless networked iBooks for branches
- Classes like HTML and Digital Cameras are demo only and are usually presented at our larger branches where attendance is higher

MOST SUCCESSFUL TECHNIQUES

- Staff: In-class exercises making them do the work: Web searches, database hunts, creating Word documents, using free e-mail services to send attachments (so staff can in turn teach the public)
- Providing staff lots of time to practice in each session has proven useful

PROUDEST ACHIEVEMENT(S)

- Developing the SJCPL Staff Training Program from its beginnings in 1997 to now, as a full-fledged department within the library system. Working with Linda Broyles, Coordinator of Networking, we created a system of classes, found it didn't work very well, retooled it and it has been successful so far. Now, we offer numerous sessions to staff, following a set of technology competencies we designed based on what other public libraries have done—Oakland, etc.

STAFF DEVELOPMENT AND TRAINING PROGRAM, PURDUE UNIVERSITY LIBRARIES

Purdue University Libraries
Staff Development and Training Program
West Lafayette, IN 47907-1530
www.lib.purdue.edu/stavdev

Contact person for the instruction program:
D. Scott Brandt
Technology Training Librarian
techman@purdue.edu

KIND OF TRAINING/TEACHING OFFERED

- Library resources, Internet, technology (Word, Windows, etc.)

MOST POPULAR CLASSES

- Library Resources: Newspaper Database Searching
- Technology: Microsoft Publisher

TO WHOM TRAINING IS OFFERED

- Staff (mostly in-house libraries staff, though some others as well)

HOW OFTEN COURSES ARE OFFERED

- Ongoing, weekly throughout the year
- Courses are scheduled prior to the beginning of a semester

SESSION SIZES

- Small (<10) and medium (10–20)
- Some demonstration sessions are large (30–50)

TYPE OF FACILITIES USED

- For technology oriented ones, a dedicated electronic classroom
- For demonstration ones, a large room with PC and projector
- For some classes which utilize videos or discussion, meeting rooms

MOST SUCCESSFUL TECHNIQUES

- Active learning: giving participants a scenario for which they have to track down information or utilize technology training (e.g., develop a Web page)
- Reflective learning: rather than an instructor answering every question, others in the class are invited to answer
- Availability of well-structured objectives and handouts in advance of the sessions to allow participants to review
- Use of online evaluation forms (easy to fill out, allows anonymous feedback)

PROUDEST ACHIEVEMENT(S)

- Development of a database registration system (K. Kielar) whereby all registrations and evaluations are entered and tracked, which can be used for performance measurement as well as instructor feedback
- Redesign of handouts to follow professional format (J. Sharkey)

PART III

SAMPLE TECHNOLOGY TRAINING MATERIALS FROM SUCCESSFUL PROGRAMS

ONE-HOUR LECTURE ON SEARCHING INDEXES

The following guide is for a 60-minute lecture/demo session on why indexes are better than search engines for finding information for papers. The outcome of the class is identification of what an index is and what kind of information can be found there. The Instructor's Guide starts off with a series of questions to generate discussion about what types of sources are needed when writing a paper and where to find them. The nature of the lecture/demo requires overhead projection of an online workstation. Participants answer questions and are engaged through discussion. The list of indexes/databases provided is an example.

INSTRUCTION GUIDE: SEARCHING A DIFFERENT KIND OF SEARCH ENGINE: INDEXES

GOAL

Demonstrate that while *Google* may be better than conventional search engines at finding "relevant" information, journal indexes provide better access to specific sources.

OUTCOME

By comparing results, students will identify indexes as reliable and "commercial-free" indexes of information suitable for term papers, reports, etc.

OBJECTIVES

Freshman students who have experience searching on the Internet will be able to:

- Identify types of information found on the Internet
- Distinguish between the inherent poor quality and incompleteness of many Web pages and sites and inherent quality and completeness of journal articles
- Describe the basic principles of relevancy ranking and page ranking
- Identify pertinent indexes from a list arranged by topic
- Compare results of searches done in search engine and index

INSTRUCTION

Start by asking students:

- "What kinds of papers or reports do you have to write?"
- "What are some of the topics?"
- "What does your instructor regard as 'term paper quality' info? What isn't?"
- "Where do you search for it?"
  - If *Yahoo!*, demo a search for the topic "NCAA gambling" *[20 matches on how to]*
  - If *Alta Vista*, demo a search for the topic "food bingeing" *[note variety; no journals]*

LECTURE/DEMO

1. What do we know about documents on the Web
   - Ill-defined domain (anything on any topic!)
   - Little structure (what's H1 for?)
   - Mostly text? Mostly images?
   - No real classification/cataloging (too much work to do by hand!)
2. What do we know about Web search engines
   - Bots do all the work... (subj directories are small compared to automated)
   - Relevance ranking/weighting most commonly used, followed by NLP, etc.
   - Few dictionaries or thesaurus (some are starting to use, but who builds?)
3. What is relevancy ranking???
   (i.e., why do most search engines give so much junk?)
   - Relevancy ranking is computer science's answer to trying to best match what you want to what the search engine database has to offer!
   - Basically it counts how many times your word (or words) appears on a page, where it appears and whether it is close to other words...
   - **IT DOES NOT KNOW WHAT YOU MEAN**
4. *Google*'s PageRank® Technology—why it seems to work so well...
   - PageRank® relies on the uniquely democratic nature of the Web by using its vast link structure as an *indicator* of an individual page's *value*. In essence, *Google* interprets a link from page A to page B as a vote, by page A, for page B
5. A good theory, but does it work?
   Try the following searches in *Google*:
   - **Subaru Forester (good—points to company first)**
   - **swollen lymph nodes (okay—points to NLM)**
   - **IRA funding (not so good—mostly financial, not Irish)**
6. Compare information on Web pages found to similar information found in journal articles [Note: prepare for this ahead of time] and discuss quality and completeness
7. Search engines that point to anything (and everything)
   Can you find term paper quality sources?
   Search in *Google* for this paper topic: biowarfare
   - **Fed of Amer Scientists (working paper)**
   - **Mother Jones (social justice)**
   - **ABC News (brief report)**
   - **Badpuppy (of an adult nature)**

- **Educate-Yourself.org (opinion)**
- **Doretk (alternative news)**

8. Indexes: Search engines that only point to newspapers and journals...
   Can you find term paper quality sources?
   AcademicSearchFulltext for topic: *biowarfare*
   - **New York Times (highly regarded news source)**
   - **Congressional Testimony (primary source**
   - **New Scientist (peer reviewed journals)**
   - **Journal of the Amer Medical Assoc (peer reviewed journals)**
   - **Public Health Reports (highly regarded information source)**
9. Conclusion: Do a preliminary search in an Internet search engine and compare the results to a search in an index (provide list of indexes by topic). *[Note: this could be an in-class exercise in which students name topics to be searched both ways, or an assignment in which students are required to turn in results from searches.]*

DATABASES BY CATEGORY

This brief guide attempts to provide access to the multiple databases the Libraries subscribes. A list of databases compiled by general category allows you to see the several different databases that may cover the same topic areas.

Agriculture/Horticulture

AGRICOLA [Wilson/Ovid]
Biological & Agricultural Index [Wilson/Ovid]
CAB Abstracts [SilverPlatter]
USGS Daily Values

Astronomy/Meteorology

Applied Science and Technology Index [Wilson/Ovid]
AAS CD-ROM Series
– Astrophysics on Disc Vol. 1-V Guide Star Catalog Computer File
AMS Conference & Symposium Preprints (1993-94)
Meteorological & Geoastrophysical Abstracts
NCDC Hourly and Fifteen-Minute Precipitation
Spatial Data Extracted from the Minerals Availability
NASA Planetary Data System

Business

Commerce Business Daily
Business Periodicals Index [Wilson/Ovid]
Business Source Elite [EBSCO]
LEXIS-NEXIS Academic Universe
Business and Industry [RDS]
Dow Jones Interactive
PROMT (Industry/Technology Sources) [InfoTrac]
TableBase [RDS]
CCH Internet NetWork [CCH]
EconLit [SilverPlatter]
Business Index
Business Newsbank
Big Business Directory
Compact Disclosure
Disclosure Select
F&S Index + Text
First Call
Hoovers Company Profiles
S&
P Stock Reports
FIS Online (Moody's)
Foreign Traders Index
National Atlas Data Bases

Chemistry

Beilstein
SciFinder Scholar
Applied Science and Technology Index [Wilson/Ovid]

Computers

PROMT [InfoTrac]
Web of Science [ISI]
Business and Industry [RDS]
TableBase [RDS]
Applied Science and Technology Index [Wilson/Ovid]
Computer Library
Computer Select
National CD-ROM Sampler: An Extension Reference
Science Citation Index
EndNote
INFO-MAC
ProCite

Crime

Social Sciences Index [Wilson/Ovid]
Social Sciences Citation Index [ISI/Web of Science]
MasterFILE Premier [EBSCO]
The Interactive Courtroom: Direct
NCJRS Document Data Base

Current issues

Business Periodicals Index [Wilson/Ovid]
MasterFILE Premier [EBSCO]
PAIS International [Ovid]
PCI Periodicals Contents Index [Chadwyck]
Lexis-Nexis
Drug Interactions Facts on Diskette

Ecology/Environment

Biological Abstracts [SilverPlatter]
Biological & Agricultural Index [Wilson/Ovid]
Applied Science and Technology Index [Wilson/Ovid]
Thrinaxodon: Digital Atlas Of The Skull
International Station Meteorological Climate
Master Weather Library (Sept. 1994)
US Navy Marine Climatic Atlas of the World (ver 1.1)
World Climate Disc: Global Climatic Change Data
Global Ecosystems Database (1992)
National Environment Watch
LandView II: Mapping of selected EPA-Regulated
PEST-BANK; Toxic Release Inventory – TRI
Hazardous Material Control & Management
Conterminous U.S. Land Cover
Characteristics Data
Southern California Bight Natural Resource Damage

Education

see also subject-specific indexes
ERIC [Ovid]
Primary Search [EBSCO]
Middle Search [EBSCO]
Integrated Postsecondary Education Data System

Employment

see also Undergrad>Collections>Reference materials> Bibliographies
Whistleblower Library and Judges' Benchbooks

OEUS
Occupational Outlook Handbook

Engineering/Physics

see also Agriculture for engineering-related topics
Compendex [Ovid]
Appl. Science & Technology [Wilson/Ovid]
Materials Sciences Collection with METADEX [Cambridge Scientific]
Vender Catalog Service [IHS]
ILI Standards Databases
Brief History Of Time, A
INSPEC

Entertainment

CAB Abstracts [SilverPlatter]
Business Periodicals Index [Wilson/Ovid]
Business Source Elite [EBSCO]
Sport Discus
From Alice to Ocean

Finance/Economics

ZACKS Investment
Business Source Elite [EBSCO]
CCH Internet NetWork
Dow Jones Interactive
EconLit
Economic Indicators [WWW GPO Access]
1994 Greenbook
Economic Analysis Tools for the Minerals Industry
National Trade Data Bank [NTDB]
Regional Economic Information System (REIS)
U.S. Global Trade Outlook 1995-2000
Working Papers received at Krannert Library
Funding Opportunities Database

General Reference

Biography and Genealogy Master Index [Gale]
Funk & Wagnalls New World Encyclopedia [EBSCO]
Reference & Directories [Lexis-Nexis]
Biographical Information [Lexis-Nexis]
Associations Unlimited
Oxford English Dictionary, 2nd Edition
The American Heritage Dictionary

Department of Defense: Telephone Directory
Vendor Master Directory
The New Grolier Multimedia Encyclopedia
GNIS: Geographic Names Information System
Magellan: full-resolution radar mosaics
US GeoData Computer File 1:1,000,000-scale DLG

Geography/Geology

Applied Science and Technology Index [Wilson/Ovid]
Compendex [Ovid]
Country Profiles [Lexis-Nexis]
Global Explorer
DeLorme Mapping (1993) US GeoData Computer File 1:2,000,000- scale DLG
ARCVIEW: Geographic Exploration System
Conterminous U.S. AVHRR. Biweekly Composites
Geophysics of North America
GEOREF
Marine Geological and Geophysical Data from the U.S. Geological Survey: Digital Data Series (DDS)
U.S. Geological Survey: USGS Open File Reports
USGS-NGIC: Geomagnetic Observatory Data
Cartographic Catalog
MapExpert
PC Globe
PC USA
STREET ATLAS USA

Government Information

GPO Access on the Web
GPO Monthly Catalog [FirstSearch]
Lexis-Nexis Academic Universe
1990 Census of Population and Housing: Summary Census of Agriculture
Comprehensive Housing Affordability Strategy (CHAS)
Consolidated Federal Funds 1984- 1993
County Business Patterns
Current Population Survey: Annual Demographic Files
Economic Census
Tiger/census tract street index CD
TIGER/Line Census Files
Congressional Record: 1985
Federal Tax Forms
American National Election Studies
CCH Access

Congressional bills online
Congressional calendar
Congressional directory
Congressional documents
Congressional record
Congressional record index
Congressional reports
Economic Indicators
Federal Register
Federally Funded Research in the U.S.
FIRMR - Federal Information Resources Management
GAO 'Blue Book' reports
Government Information Locator Service (GILS)
History of Bills and Resolutions Database
OnPoint
Privacy Act Issuances
Public Laws of the 104th Congress
U. S. Code
Unified Agenda
United States Code
United States Government Manual

Health and Medicine

Health Source Plus/Advice for the Patient [Inspire]
CINAHL [Ovid]
MEDLINE [Ovid or Inspire]
Social Sciences Index [Wilson/Ovid]
EconLit [SilverPlatter]
Health Topics/Medical Abstracts [Lexis-Nexis]
The Longitudinal Study of Aging
A.D.A.M. Standard
Food Science and Technology Abstracts [FSTA]
Food/Analyst Plus
CDP File - National Center for Chronic Disease
National Health Interview Survey
Human Nutrition
International Pharmaceutical Abstracts (IPA)
Micromedex CCIS
PsycINFO
PsycLIT

Humanities

Humanities Index [Wilson/Ovid]
Arts and Humanities
Social Sciences Citation Index [ISI/Web of Science]

PCI Periodicals Contents Index [Chadwyck]
Literary Resource Center [Gale]
Biography and Genealogy Master Index [Gale]
Linguistics and Language Behavior Abstracts (LLBA) [Soci]
MLA Bibliography [Ovid]
Reference Manager
Books In Print [SilverPlatter]
ARTFL [U of C]
KEY MEGA CLIP ART
Arts and Humanities Citation Index
Contemporary Authors
Dictionary of Literary Biography
French National Biography
Heracles
Perseus
Contemporary Literary Criticism Select
ADMYTE Spanish Medieval Texts
Books In Print®
LIZ Letteratura Italiana Zanichelli
MLA International Bibliography

International

PAIS International [Ovid]
Lexis-Nexis
U.S. Exports History
U.S. Exports of Merchandise
U.S. Foreign Affairs on CD-ROM 1990-1995
U.S. Imports History
World Development Report

Legal

Federal Register, Public Laws & United States Code [GPO]
Lexis-Nexis
CCH Internet Tax NetWork [CCH]

Mathematics

see also Engineering topics
MathSciNet [AMS]
Appl. Science & Technology [Wilson/Ovid]
MathSci Disc

Newspapers

Newspaper Source [EBSCO]
Dow Jones Interactive
Lexis-Nexis

Resources>Virtual Library>News or off Undergrad Library>Collections page

Politics

see also Government topics
Lexis-Nexis
PAIS International [Ovid]
Social Sciences Index [Wilson/Ovid]
Congressional Directory & US Government Manual [GPO]
ABC Pol Sci on Disc

Social Sciences

Social Sciences Index [Wilson/Ovid]
PAIS International [Ovid]
Business Periodicals Index [Wilson/Ovid]
Humanities Index [Wilson/Ovid]
Social Sciences Citation Index [ISI/Web of Science]
Ethics CD
Ethnic NewsWatch
America: History and Life
Global Historical Fields (1994)
The First Emperor Of China

Standards

ILI Standards [UK Web-based] for ordering
U.S. Patent Citation Database [COS]
Vender Catalog Service [IHS]
CASSIS
DoD Standardization Directory
Worldwide Standards Directory

Statistics

CCH Internet Tax NetWork [CCH]
Agricultural Statistics 1994
American Housing Survey
Catalog of electronic Data Products
Compustat
County & City Data Book: 1988
Datastream
High School & Beyond
International Financial Statistics (IFS)
National Center for Education Statistics
National Post-Secondary Student Aid and Study: SASS - Schools and Staffing Survey

Statistical Abstract of the U.S.
Statistical Masterfile
Statistical Universe
Traffic Safety Data 1988-1993
Trends in Health in Older Americans, U.S.
Truck Inventory & Use Survey (TIUS)
U.N. Statistical Yearbook
U.S. Imports of Merchandise

Technology

see also Engineering or Agriculture for related topics
Appl. Science & Technology [Wilson/Ovid]
Predicasts PROMT [InfoTrac]
Food Science and Technology Abstracts [SilverPlatter]
Business and Industry [RDS]
ArticleFirst [FirstSearch]
Dictionary of American Naval Aviation Squadrons

HANDOUTS FOR TRAINING SESSION

The following set of handouts is for a 90-minute lecture/demo session on the EBSCOHost index and databases (also called *INSPIRE*, in Indiana). Since the outcome of the 90-minute class is comprehension of terminology and features, participation follows a "watch and do" approach (i.e., presented in an electronic classroom environment where participants are sitting at workstations). The Instructor's Guide shows which terminology is defined and which features are demonstrated. Participants practice the same search, and then variations, along with the instructor. The instructor checks for comprehension by quizzing participants ("How do we browse the subject index listing?") and checking performance along the way ("Did everyone get 21 hits?").

INSPIRE/EBSCOHOST—INSTRUCTOR'S GUIDE

PREREQUISITES

Netscape, or experience using a Web browser
Familiarity with searching for journal articles

PRELIMINARY

What is an index, what are databases and why would someone use them?

Indexes are like Access—the program that allows you to search for data and information. A database is structured table of content. In the library world, indexes are the primary resources used to find journal articles and similar materials.

What's the difference between Inspire and other indexes like Ovid or WilsonWeb?

Each index has a different interface or "look and feel" though many have similar features for searching. Inspire's databases are aimed to be accessed by users in public libraries and in a variety of education environments—elementary school, high school, college.

OBJECTIVES

1. Learners will be able to identify Inspire/EBSCOhost databases by topic
   - **Business:** Business Source Premier
   - **Education:** ERIC, Professional Development Collection
   - **General:** EBSCO Animals, Funk & Wagnalls New World Encyclopedia, MAS FullTEXT Ultra, MasterFILE Premier, Newspaper Source
   - **Health/medical:** Clinical Reference Systems, Comp MEDLINE w/ MeSH, Health Source Plus, USP DI Volume II, Advice for the Patient
   - **Primary/secondary school:** EBSCO Animals, Funk & Wagnalls New World Encyclopedia, Middle Search Plus, MAS FullTEXT Ultra
   - **Research:** Academic Search Elite, Business Source Premier, ERIC, Comp MEDLINE w/ MeSH, Professional Development Collection
2. Learners will be able to identify five types of searches within the EBSCO interface

- Subject, Journal, Keyword, Advanced, Expert, and Natural Language

3. Learners will be able to search by keyword
    - ~ Internet addiction without any Boolean or limits
   - Identify Benefits of a keyword search
   - Identify limit options: Limit to Date, Publication Type, Journals, Full Text or Peer Reviewed
     - ~ Internet addiction with limits of date and peer reviewed
4. Learners will be able to conduct an advanced search
   - Search multiple terms by combining search sets
     - ~ Set 1: *Internet addiction AND Set 2: treatment*
   - Identify limit options from the Advanced Search screen
5. Learners will be able to search phrases and multiple terms using Boolean/Proximity operators, truncation, and field codes
    - ~ These search tools can be used in any of the search screens
    - ~ Advanced search: *internet addict* AND treat**
    - ~ Keyword: *internet NEAR addiction*
    - ~ Expert search: *TI internet addiction*
    - ~ Expert search: *S1 AND S2*
6. Learners will be able to conduct a natural language search
    - ~ *treatment of internet addiction*
7. Learners will be able to browse journal titles or subject headings
   - Identify benefits of each search method
   - Select items to view records or documents
     - ~ Search through alphabetical lists and select items to view records/documents
     - ~ Journal: *Harvard Medical News Report*
     - ~ Subject: *internet addiction*
       - Identifies how many resources are associated with the subject
       - If you select subject specifically, it also shows SEE ALSO links
8. Learners will be able to display search results
   - Identify various formats on a search results list
     - ~ Citation and abstract only
     - ~ Full Text (as HTML)
     - ~ Full Page Image
     - ~ Check Linked Full Text
9. Learners will be able to Print/Save/Email results
   - Identify output options for delivering a results list or full text

- ~ For Print/Save, after clicking the Submit button use the browser functions
- ~ Email arrives with the From line saying: ***ehost@epnet.com***

10. Learners will be able to locate alternative access and identify 2 additional resources
   - Identify hyperlink on Indexes page
     - ~ As noted on handout, www.inspire-indiana.net/ leads to another version of Inspire which was designed for "All citizens of Indiana"
     - ~ It can also be accessed from the Inspire databases hyperlink on the Indexes page
   - Identify 2 additional resources
     - ~ NetFirst is a database of library quality URLs
       - – FYI - NetFirst is now available through the FirstSearch interface
     - ~ Other Reference Sources points to local/regional sites

Inspire stands for **IN**diana **SP**ectrum of **I**nformation **RE**sources. This is a statewide initiative to provide free access to electronic resources to everyone in Indiana. Inspire's databases are aimed toward users of all types of libraries. EBSCOhost is the search software and interface to search the databases. This resource is known by both names, EBSCO or Inspire. Many of the resources within the Inspire databases are full text articles. Listed below are the databases categorized by general subject area.

- **Business**: Business Source Premier
- **Education**: ERIC, Professional Development Collection
- **General**: EBSCO Animals, Funk & Wagnalls New World Encyclopedia, MAS FullTEXT Ultra, MasterFILE Premier, Newspaper Source
- **Health/medical**: Clinical Reference Systems, Comp MEDLINE w/ MeSH, Health Source Plus, USP DI Volume II, Advice for the Patient
- **Primary/secondary school**: EBSCO Animals, Funk & Wagnalls New World Encyclopedia, Middle Search Plus, MAS FullTEXT Ultra
- **Research**: Academic Search Elite, Business Source Premier, ERIC, Comp MEDLINE w/ MeSH, Professional Development Collection

SELECTING A DATABASE

After clicking a database link from the Indexes page, you will need to choose the database(s) you want to search.

1. Click the checkbox next to the database name to select it
2. Click the Enter button to begin searching

☑ **Academic Search Elite**
Provides full text for over 1,250 journals covering t
here for a complete title list. Click here for more in

☑ **Newspaper Source**
Provides selected full text articles from 143 U.S. an

☑ **Health Source Plus**

You are automatically taken to the keyword search screen. However, from the navigation bar you can choose the type of search you want to conduct.

KEYWORD SEARCH

Keyword Search

The search terms entered will be searched in all fields including abstracts and full text. It provides the broadest possible search. The keyword search provides some limits. You may limit the search to only the full text, a specific magazine, and date published. You may expand the search to search for related words that are synonyms and plurals of search term(s).

Databases: *Academic Search Elite; Newspaper Source; Health Source Plus*

Find:

Search

Clear

Enter keywords you wish to find. You may separate keywords with **and, or,** or **not**. For search examples, see Search Tips.

Limit Your Results:

☐ Full Text

Magazine:

Date Published: Month Yr: to Month Yr:

Expand Your Search:

☐ also search for related words

☐ search within full text articles

ADVANCED SEARCH

Advanced Search

The concept behind Advanced Search allows you to group like terms together and connect them with other terms to create a more focused search. Advanced search allows you to select specific fields in which to limit the search terms, much like assisted search in the Libraries Catalog.

1. In the Find: text box, type like search terms together
2. Limit to specific field by using the Field drop down menu
3. Select a Boolean operator to connect the search sets together
4. In the next search text box, type the next set of search terms

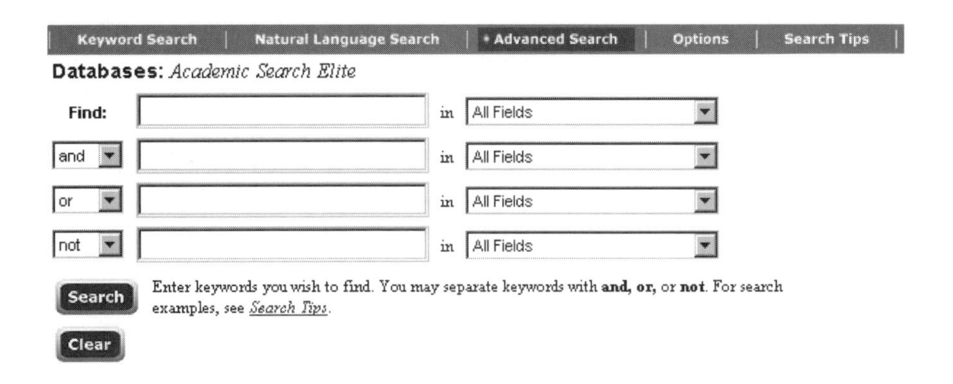

SEARCH OPTIONS

Boolean	Proximity	Wildcards	Field Codes
AND OR NOT	Near (N) Within (W#)	* Truncate ? single character	Subject SU Title TI Author AU
(cat OR feline) AND travel*	tax N2 reform Web W3 security	comput* wom?n	SU computers TI time AU gates

NATURAL LANGUAGE SEARCH

Natural Language Search

Natural Language searching allows you to type search terms just as you would say them in natural speech. Relevancy Ranking is used to build the search results. The more words that appear in an article, the more relevant the record is and the closer to the top of the result list it will appear.

1. Click the Natural Search Language button to display the search screen
2. Type your search phrase in the Find: text box
3. Set your limits, if desired
4. Click the Search button

Keyword Search | Natural Language Search | Advanced Search | Options | Search Tips

Databases: *Academic Search Elite*

Find: [ ] Search

Clear

For a *Natural Language* search, type a phrase or sentence which describes what you are looking for. Quoted phrases or keywords will always be included in your results. For search examples, see Search Tips.

Limit Your Search:

☐ Full Text

Publication Type: [ ]

Journal: [ ]

Date Published: Month Yr: [ ] to Month Yr: [ ]

☐ Peer Reviewed

EXPERT SEARCH

Expert

The Expert Search uses keyword search, search history, and special limiters to help you focus your search. You can combine searches listed in the search history panel as well as use special field codes to do a field limit. The field codes differ depending on the databases you are searching.

1. Click the Expert button
2. Type in search terms using field codes and Boolean as needed or combine search sets using the number of the search and Boolean
3. Limit with the options provided
4. Click the Search button
5. View records as you normally would
6. Return to the Search History by clicking the Expert button

Find: TI internet addiction Search

Clear

Enter keywords, search history ID numbers, or perform a command line search. For databases with Medical Subject Headings (MeSH), click on the Browse MeSH icon button above. See Search Tips.

Go to: Limiters **Show:** Field Codes

Your Search History:

#	Query	Limiters	Results	Revise
S1	internet addiction			

Limit Your Search:

☐ Full Text	Magazine: [ ]
Date Published: Month Yr: [ ] to Month Yr: [ ]	☐ Peer Reviewed
Publication Type: Periodical, Newspaper, Book	Number Of Pages: [ ]
☐ Cover Story	Articles With Images: Full Page Image, Text with Graphic

SUBJECT SEARCH 

Subject headings assigned to a resource reflect the content and focus of the resource. When doing a Subject search, you can browse a list and search for a specific subject heading. Each heading will have the number and type of resources indexed with that heading

1. Click the Subject Search button
2. Select a category (All is the default)
3. Type a search term or use the up and down buttons to browse the list
4. Click the hyperlink to display the resources with the specific subject heading

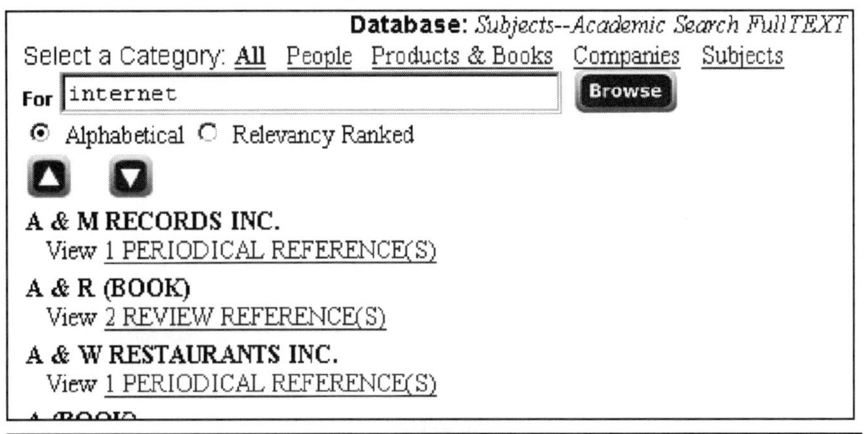

JOURNAL SEARCH 

You can use this option to see if a specific journal is indexed in a specific database. As with a Subject Search, you can browse a list and search for a specific journal title.

Database: *Journal Name-Academic Search FullTEXT*

For [ ] Browse

Alphabetical Relevancy Ranked

Search Select one or more items and click Search!

- ABA Banking Journal
- ABA Journal
- Abstract Machine Models for Parallel & Distributed Computing

DISPLAYING SEARCH RESULTS

When you conduct any search, the search results are automatically displayed in reverse chronological order. Any resource that is available in full text will be marked with one of three full text icons.

The Full Text icon indicates that full text is available in an HTML file

The Full Page Image icon indicates that the article is a pdf file of the original article as it appeared in the resource.

The Check Linked Full Text link is present when the resource is not in the other formats but is available in another EBSCOhost database or electronic journal.

Click the icon to display the full text of the document in either HTML or as a pdf file. The Full Text in HTML lists the citation, subject headings, and navigation between the results list and the full record. Not every record has full text available or both formats available.

EBSCO HOST | New Search | Company Directory | Image Collections | Choose Databases | Online Help

5.	☐	New on the Net. PC World, Feb2001, Vol. 19 Issue 2, p41, 1/5p Full Page Image
6.	☐	Honey, I Shrunk the Bits! Removable Storage Gets (Really) Small. PC World, Feb2001, Vol. 19 Issue 2, p56, 1p Full Page Image
7.	☐	A Colorful Compaq Home System With Kick. PC World, Feb2001, Vol. 19 Issue 2, p76, 1p Full Page Image

PRINTING, E-MAILING, AND SAVING

Print/E-mail/Save

Just like many other electronic databases, the searcher has the option to print, e-mail, or save the list of search results, a marked search list, or full text. To print or save a pdf file, use the print & save options in Acrobat Reader. To print, e-mail, or save an HTML file, follow the steps below.

1. To initiate any of these options, click on the Print/E-mail/ Save button.
2. Select the 'What do you want print, e-mail, or save?' option desired
3. Indicate the information format
4. Select Yes or No to include or not include full text
5. To include or not include page images, select Yes or No
6. Then select the delivery format: print, save, or e-mail
   - When you select the Save or Print option, the search list or full text will be redisplayed as a plain HTML file. Use the Print or Save options of your Web browser.
   - When e-mailing a search or full text, type in your e-mail address in the available text boxes.
7. Click the Submit button Submit

Note: All items marked on the result list will remain marked until a new search is performed via New Search or Refine Search.

1. What do you want to print, e-mail or save?	Only marked items Entire result list
2. For all the items, show the following.	Detailed info (with abstract, if available) Bibliographic manager format Links to marked items.
3. If available, also include the full text?	No Yes Highlight the search term(s) in full text
4. If available, e-mail Page Images?	No Yes - Each Article will be e-mailed separately
5. How should the items be delivered?	Display to **Print** SUBMIT and then use the browser Print option Display to **Save** SUBMIT and then use the browser Save option Via **e-mail** - contains text only; graphics embedded in full text will not be included Address: Subject:

THE OTHER INSPIRE INTERFACE

For users who want to access and search these various databases, the State of Indiana provides its own interface as well as a link to the EBSCOhost interface. The non-EBSCOhost interface provides links to the same databases as well as access to NetFirst, a database of library quality URLs, and other quality resources elsewhere on the Internet, including any sites of local interest.

ACCESSING THE STATE INTERFACE

1. Type the URL www.inspire-indiana.net/ in the location bar of your Web browser
2. Click on "GO" on the Start Here form GO!
3. Click the Inspire interface link
4. Select the database or topic area to search

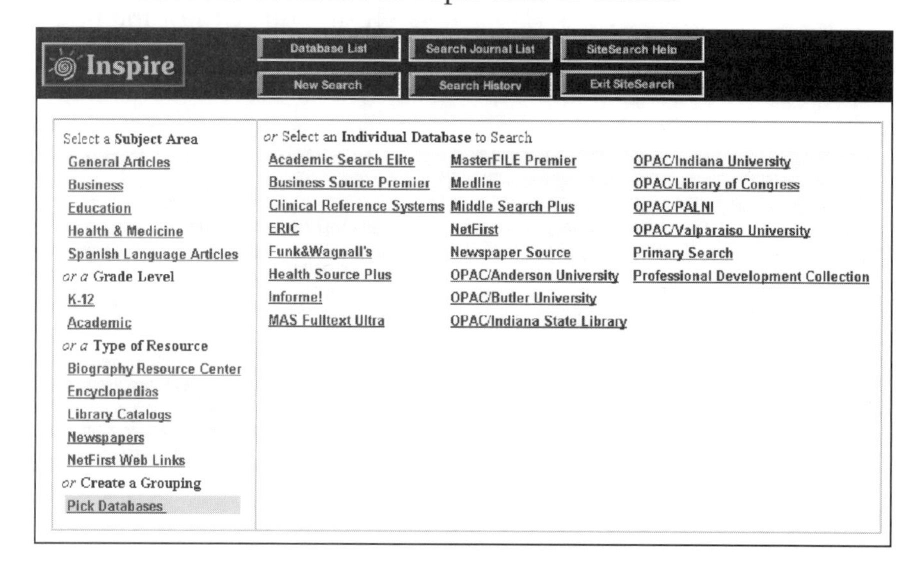

INSPIRE/EBSCOHOST PRACTICE QUESTIONS

1) When using Medline, the EBSCO interface doesn't have the map to keywords to subject headings feature. How are you supposed to know if you are using the correct terms when searching something like heart attack?

2) A student at Purdue University wants to major in biomedical engineering and minor in management, with the prospect of taking over her father's business, which manufactures artificial joints. In particular, she wants to know how that industry is doing these days. Which Inspire indexes might she choose, and what might her search statement be?

3) An education major is looking for "chemistry resources for teaching on the Internet"—and says she tried Yahoo and Alta Vista and just got frustrated. How could the Inspire databases help her?

4) Is there an online book review of Donna Tartt's *The Secret History* in Inspire?

5) Someone's daughter just got a macaw, which is evidently some kind of parrot, and wants to know if her parents will baby-sit it over the upcoming semester break. They think they need to read up on this before they commit to anything—where might they look for some information?

6) A student heard on the radio that gingko is supposed to help with short-term memory retention, and feels like he could sure use some help there. But, all the student knows about gingko is that the fruit from the gingko tree smells horrible —was the radio reporter talking about the same thing?

Inspire/EBSCOHost Course Evaluation					
Put a check mark in the column which best describes your experience…					
	Strongly Agree	**Agree**	**N/A**	**Disagree**	**Strongly Disagree**
This course has clearly stated objectives.					
I am able to locate Inspire/EBSCOhost databases					
I can select single or multiple databases to search on.					
I can identify Inspire/EBSCOhost data-bases by topic					
I can search by keyword for a topic in various databases					
I can search multiple terms by combining search sets					
I can print, save, and email records & results					
I can browse by journal titles or subjects					
I can conduct a natural language search					
I can locate the "other" Inspire interface					
The instructor speaks clearly, gives examples, slows down and repeats as necessary.					
The instructor asks questions and listens to answers, and allows me to ask questions and gives helpful responses.					
Strategies such as calling for participation, asking students to demonstrate skills and walking through examples are used by the instructor to help increase learning					
The facilities for this course are excellent.					
I highly recommend this course for the content covered.					
I highly recommend this course for the way it is taught.					
I would enjoy taking another class which utilizes the learning techniques and qualities as noted above.					

HANDOUTS FOR A WORKSHOP ON DESIGNING INSTRUCTION

The following set of handouts is for a three-hour workshop introduction to the design of a module of instruction. Since the outcome of the class is to have a framework upon which to design and develop instruction, participants work through examples for reinforcement. The objectives include intellectual behavior and problem-solving skills, and participation follows a "lecture and do" approach (in this case presented in an environment with tables and chairs). For instance, after discussing four-part objectives, participants practice creating them. The Instructor's Outline details the objectives and exercises practiced. Participants work in groups to apply techniques to case studies, with the instructor checking in on progress.

DESIGNING INSTRUCTION WORKSHOP

INSTRUCTOR OUTLINE

Obj 0	Given course handouts (**C**), identify (**B**) general objectives to be covered and determine applicability to Internet/Technology training. (**D**)		
Type	Verbal & Intellectual	**Time**	9:00-9:15 (15 mins)
Prereq	1) Organizational skills; 2) Interest in training; 3) Describe training requirements		
Strategy	1) Survey users regarding prereqs and display answers on flip chart; 2) Review PPT slides on objectives and ask if they know how they pertain to them; 3) Describe/demo context		
Obj 1	When analyzing a situation to create objectives (**C**), describe 3 aspects (**B**) of learners/learning—learner needs (current, future and gap), learning preferences(verbal, auditory or tactile and deductive or inductive), level of skill (beginner vs. advanced). (**D**)		
Type	Verbal & Intellectual	**Time**	9:15-9:40 (25 mins)
Prereq	1) Can define who their learners are; 2) Acknowledge differences among learners		
Strategy	1) Parody comparing personality to learning; 2) Define & demo learner needs & gap analysis, learning preferences, level of skill; 3) Apply aspects to their learners by discussion		
Obj 2	When developing (4-part) learning objectives(**C**), determine (**B**) audience, behavior, condition, and degree to develop (**B**) a complete objective with outcome. (**D**)		
Type	Verbal & Intellectual	**Time**	9:40-10:10 (30 mins)
Prereq	1) Can identify "measurable verbs"; 2) Can define "demonstrable outcome" and "degree"		
Strategy	1) Define ABCD; 2) Show example; 3) Group exercise building objectives		
Obj 3	Given a list of terms and definitions (**C**), correctly (**D**) identify and distinguish (**B**) between 5 types of learning objectives.		
Type	Verbal	**Time**	10:10-10:30 (20 mins)
Prereq	1) Can identify learning objectives; 2) Can relate personal experiences to different types		
Strategy	1) Define 5 types (affective, motor, verbal, intellectual, cognitive); 2) Review the handout; 3) Quiz slide: given objectives, determine type		

Type	BREAK - BREAK - BREAK	**Time**	10:30-10:45 (15 mins)
Obj 4	Given a list of terms and definitions (**C**), identify and distinguish (**B**) between several types of strategies (**D**)		
Type	Verbal	**Time**	10:45-11:05 (20 mins)
Prereq	1) Define instructional strategy; 2) Can relate personal experiences to different types		
Strategy	1) Review types handout; 2) Quiz slide: given objectives, determine type		
Obj 5	When presented with a list of examples (**C**), match objectives to strategies(**B**) to ensure learner perspective is accounted for in module of instruction. (**D**)		
Type	Cognitive	**Time**	11:05-11:20 (15 mins)
Prereq	1) Can identify learning objectives & strategy;		
Strategy	1) Review matching handout; 2) Exercise		
Obj 6	Apply objectives and strategies to Internet-related instruction		
Type	Intellectual & Cognitive	**Time**	11:20-11:45 (25 mins)
Prereq	1) Can identify learning objectives & strategy; 2) Can relate personal Internet experiences		
Strategy	1) Demo a module, show step-by-step where/how objectives are applied; 2) Group work reviewing a scenario and determining learner/learning, objectives, strategies for it		
Type	WRAP-UP & EVALUATION	**Time**	11:45-12:00 (15 mins)

TERMINOLOGY AND DEFINITIONS

Learner Will Be Able To...

1. Describe assessment of learners/learning
2. Account for differences in learning preferences
3. Accommodate differences in learning preferences
4. Develop four-part learning objectives
5. Identify five types of learning objectives
6. Define instructional strategies

1. Assess Learners Needs
   Assessment is like the reference interview—you have to figure out what people really want/need! Define the problem clearly. Then check for...
   - What they don't know (gap)
   - Where they apply learning
   - What outcomes are needed
2. Account for Preferences
   - **Auditory**: needs to hear instructor and others speak for reinforcement
   - **Visual:** prefers to visualize through images, handouts, etc.
   - **Tactile**: needs hands-on, work sheets to reinforce
   - **Deductive**: explain concept and rule and they figure out application
   - **Inductive**: use examples or problems to work up to concept through application
3. Accommodate levels with variety of resources and materials
   Beginner needs: gentle structure, background, time to reflect
   Advanced: recipes and tips, to the point
   Everyone complains about mixed levels of experience/skill in a class:
   Post objectives prior to allow self-assess
   Arrange "needy" by "givers"
   Get steely—don't sacrifice the class
4. Objectives and Outcomes (Heart of Instruction)
   Audience: level, type of learner
   Behavior: what specific action
   Condition: under what circumstances
   Degree: how often, to what degree
   To yield a measurable **Outcome**
   Audience: Beginning searchers

Behavior: will be able to generate 2 BT and NT terms for their topic
Condition: when doing a search in the Yahoo search directory
Degree: which will result in at least two useful categories
Outcome: Categories to look under

5. Types of Learning/Objectives
   Affective: related to emotion
   Motor skills: physical dexterity
   Verbal knowledge: memorization
   Intellectual skills: making decision by choices or options
   Problem solving: apply, determine, figure out
6. Instructional Strategies
   Matching a type of objective to a method to achieve the outcome
   For instance: using a quiz to reinforce a lecture on terminology

CASE STUDIES

Case Study #1

A chemistry librarian in a university setting is doing an exercise in evaluating Web sites for sophomores. She has specific chemical compounds that she wants students to look up on the Web, and she wants them to evaluate which is better and why.

What are the outcomes she is trying to achieve?

What should be the primary objective(s)? Secondary?

Which type(s) of objective(s) are involved? What are some possible strategies (formats, exercises, materials) to use?

Answer: She might decide that <u>using and comparing</u> the resources was more important than revisiting evaluation skills, which had been covered in the previous year. Thus, she would focus on a handout that listed strengths and weaknesses of each.

Case Study #2

A music instructor in a community college setting is required under local directives to send first year students to a "library course on searching for information." The students are not required to write a paper for this class, but she wants it to be meaningful and related to her course.

What are the outcomes she might try to achieve?

What should be the primary objective(s)? Secondary?

Which type(s) of objective(s) are involved? What are some possible strategies (formats, exercises, materials) to use?

Hint: After consultation with a librarian, she might realize that a "hunt" was out of the question (i.e., find out where Handel is buried) and that actually writing a paper (i.e., defining a topic) was too much. She might decide that the important outcome was to emphasize basic searching skills. She might settle on an exercise in which students had to find and compare a library-based resource and a Web-based resource, listing both objective and subjective criteria for using them to find interesting facts related to specific periods in music history.

Case Study #3

A training librarian for a public library is deluged with requests from patrons for a class on how to use e-mail. Her library's most popular class is a hands-on Internet Basics course which covers navigational tools, the library subject list and basic search techniques.

What are the outcomes she might try to achieve?

What should be the primary objective(s)? Secondary?

Which type(s) of objective(s) are involved? What are some possible strategies (formats, exercises, materials) to use?

Hint: You might decide on basic skills in sending/receiving messages (but also had to initiate and account). "Introduction to Web-based E-mail: Free WWW-based e-mail is for anyone—anywhere. Join this hands-on introductory class to get an e-mail account and learn how to use it!" You might discuss addressing, how e-mail works and begin sending messages. Mousing skills strongly recommended, or participation in the SJCPL Hands-on Internet classes. Pre-registration could be required. Class size should be limited. This class could be repeated to allow for more participants. (Might be followed by Making the Most of E-mail: Make the most of your e-mail account! Learn how to send attachments, use your address book, and other special features of e-mail!

WEBLIOGRAPHY

Teaching and Libraries

University of Ulster Library Training Page—specifically related to libraries
www.ulst.ac.uk/library/training/

Librarians' Index to the Internet—Internet training
www.lii.org/search?searchtype=subject;query=Internet+training

Internet/Searching

Using Search Engines Better
www.searchenginewatch.com/facts/index.html

Technology Toolkit from The Node (with links)
http://thenode.canlearn.ca/techtoolkit.shtml

Internet Search Strategies for Teachers (with links to other sites)
www.2learn.ca/research/rss.html

Web Search Strategies (includes Researching Companies Online tutorial)
http://home.sprintmail.com/~debflanagan/main.html

Tutorial: Guide to Effective Searching of the Internet—insight from a producer's point of view
www.completeplanet.com/Tutorials/Search/index.asp

Fun with and about technology—quizzes, games, etc.
www.quia.com/dir/tech/

Learning and Technology

Learning Technology Web resources
www.keele.ac.uk/depts/cs/Stephen_Bostock/keywords/index.html

Distance Learning Links For Librarians
www.shu.ac.uk/services/lc/dl/index.html

Extensive collection of links to Web-based training sites from Brandon Hall
www.multimediatraining.com/links.html

Other

Personality Test
http://users.rcn.com/zang.interport/personality.html

SYLLABUS FOR INFORMATION LITERACY COURSE

The following set of handouts is for an eight-week (16 hours) introduction to information literacy skills, specifically aimed at students in an electrical engineering technology program. The outcome of the course is for freshmen to use indexes as a primary source for information supporting coursework and develop sound information seeking skills. The Syllabus notes various policies regarding the course, and outlines the class schedule. The Course Guide details content covered in each session. Homework and guide for the Final Project are included as well. The Final Project is a report that serves as an annotated bibliography for a topic of their choosing. Basically the course follows a format in which students identify skills that are needed as they research their topic (e.g., when asked, "Where do you search for background information," a discussion eventually points toward books and searching the online catalog). Various sessions follow a "demonstration and do" approach.

SYLLABUS

SYLLABUS for Information Strategies: An Information Literacy Aimed at Students in an Electrical Engineering Technology Program

COURSE PURPOSE

This course introduces you to the core concepts of information retrieval and essential techniques for finding, analyzing, organizing, and presenting information within a simulated company research environment. Much of the coursework will be based around this Information Strategies Process:

1. Define problem or topic
2. Decide where and how to search
3. Search appropriate resources
4. Evaluate what was found

5. Review and revise search
6. Organize information

LEARNING GOALS

Develop information literacy skills for EET students at Purdue University
Approach research and instruction from a problem-solving format
Apply learning to real-life research situations
Create an electronic portfolio of the process for self and group assessment

LEARNING OUTCOMES

Upon successful completion of this course each student will be able to:

Use information to identify and solve problems
Build and analyze information retrieval strategies for research
Differentiate between types of resources
Apply key concepts to retrieve information
Evaluate and organize information

COURSE POLICIES

Grading Policy:

This course will be graded as A (90 percent or higher), B (80–89 percent), C (70–79 percent), D (60–69 percent), F (less than 60 percent). Your grade will be determined by the following:
A student's final grade is based on his or her individual work and group performance.
The individual grade is based on class participation, homework, and performance in the presentation of group's final project.
The group grade is based on the written portion of the final project, and peer assessment and review of group work.
An EET student receiving an F will be required to repeat this course.

Participation and Attendance:

Your attendance is required

1. The nature of the course is cumulative. The in-class and homework assignments are designed to build upon each other. It will be important that you complete all of them on time.

2. A portion of each class is devoted to group activities and planning.
3. Attendance will be taken, and there will be in-class assignments which will be collected to count toward the class & group grade.

Class participation:
Physically present for all classes
Individual contributions in class
Contributions to group work

If you must miss a class due to an emergency, you must contact one of the professors as soon as possible as some additional assistance will need to be scheduled.

Cheating:

Scholastic dishonesty (cheating) is not tolerated at Purdue. It can be generally defined as giving or receiving aid in examinations or on assignments which are intended to be done individually, or the presentation of the work of other persons as one's own. The only copying or sharing of work permissible is between members of an assigned class team, and then only for team projects. The individual grade is based on class participation, peer assessment, and performance in the presentation of group's final project.
The group grade is based on the written portion of the final project, and peer assessment and review of group work.

Homework Assignments:

All homework assignments are due on time.
Group Review:
You will develop a project for research at the beginning of the course. This will be a group interactive project, which will culminate into a final presentation. All groups will coordinate their work.
Final Project.
You will work as an individual but coordinate with your group to complete the final project.
The final project will be due at the first class of the last week.
Groups will give a ten-minute presentation of their final project.
The final project is designed to assist you to achieve the course objectives. Please review them when you work on the project.
This project is worth nearly half of the course grade and shall be assigned an individual grade based on its overall quality. In the final week of class, each group will give a ten-minute presentation on their final project. Each group member is expected to be

a productive member of the group. A complete description of the contents and format for the final project can be found on this page.

CLASS SESSION OUTLINE

Session One
Introduction to Simulated Company Research
  Determine a research problem for a company setting
Session Two
Defining a topic
  Construct a statement of the problem
  Formulate possible solutions/hypotheses
  Identify types of information needed to solve the problem
Session Three
Where do I go for the best information?
  Identify specific types of information sources
  Differentiate between types of resources
  Homework One Due:
  Proposed Topic For Research
Session Four
How do I find relevant resources?
  Apply various methods of searching
  Demonstrate how to retrieve information
Session Five
Is the Internet a scholarly resource?
  Use computerized resources to locate sources of information
  Identify the nature of information
  Homework Two Due:
  Revised Topic For Research
Session Six
Now that I found useful information, what do I do?
  Create a working bibliography using the MLA format
  Develop a Web page using HTML
Session Seven
How do I really know if my information is good enough?
  Identify the elements that determine the appropriateness or authority of a source
  Evaluate the quality of the information
  Homework Three Due:
  Background Sources for Research
Session Eight
Information overload!!! Enough is enough—how do I organize it?
  Recognize the purpose and structure of disseminating information

Organize information into a Web-based portfolio

Session Nine

I have a lot of information. How will I use it effectively?

Revise the information retrieval strategies as needed to improve the research

Homework Four Due:

In-depth Sources for Research

Session Ten

Am I going in the right direction?

In-class review of progress

Session Eleven

More searching?

Use help pages to identify advanced searching techniques

Formulate searches using Boolean and proximity operators

Compare results using basic vs. advanced searching techniques

Homework Five Due:

Up-to-date Sources for Research

Session Twelve

Am I there yet?

Identify relevance or precision of search results

Increase relevance/precision of results using advanced techniques

Session Thirteen

What about information specific to my field? Like—patents?

Describe the difference between specialized and general resources

Find three types of specialized resources

Session Fourteen

So, how does this course really affect my life?

Identify three key sources from your final project

Extrapolate two things learned while searching

Create an outline for the presentation

Session Fifteen & Sixteen

Presentations

COURSE GUIDE

SESSION ONE

Fill out cards (Name, e-mail, "What you would like to learn...") & survey

Background on course ("What have you heard?"/historical perspective)

Overview: pseudo-company setting, research problem, Basic Process...

Discussion of SC-179 as learning environment
(Need career account; log-in only when lab portion starts; speak up!)

SESSION TWO

Defining a topic: Demonstration & Example [computers]
Construct a statement of the problem with supporting conditions
Formulate possible solutions/hypotheses
Identify types of information needed to solve the problem
Netscape tutorial: emphasize Preferences (fonts/mail/cache) & Bookmarks in-class exercise
KWL Group work
In-class exercise

SESSION THREE

Opening Discussion: interview with Andreesen/Berners-Lee/Cerf (www.computerworld.com/home/features.nsf/all/990104forecast)

- What do you think of when you hear the phrase 'bad information'? Can anyone 'rebut or replace' it?
- If there is 'bad' information, is there 'good' information? Is there 'best' information? What would it be?
- In what ways could technology be used to 'warn about mis-information'? Effective or not?
- What dangers are you aware of regarding info on the 'Net? In what way is the Web a 'breeding ground'?
- What are the trade-offs for having a 'fluid and highly connected' information network?

Where do you go for the "best" information? *[Timeline...]*
Identify specific types of information sources
Differentiate between types of resources
KWL Group work
Repeat in-class exercise
Homework One Due: Proposed Topic For Research

SESSION FOUR

How do I find relevant resources?
Last time we talked about "good" information... how do we go about finding it? or resources?
(different types & where they might be; Internet vs. library)
(kinds of tools to use)
Apply various methods of searching

Simple keyword
Controlled vocabulary
  1) subject headings
  2) BT/RT/NT structure
  3) mapping/auto cross ref
Boolean operators
  1) AND
  2) OR
  3) NOT
Proximity operators
  1) ADJ
  2) NEAR
  3) SAME

BT
Broader Term
RT
Related Term

NT
Narrower Term

Field limiting—lang, date,
Relevance ranking—db ratio; occurrence; their weight; your weight
Demonstrate how to retrieve information

- LookSmart
- AskJeeves
- Webcrawler
- THOR...

SESSION FIVE

What did we cover last time? (searching) What stood out for you? We talked last week about different kinds of information, including what is reliable or authoritative... Is the Internet a reliable or authoritative resource? Can you get reliable or authoritative sources on the 'Net?

- Internet sources vs. Internet as delivery for sources

Use computerized resources to locate sources of information
  www.lib.purdue.edu/~techman/undernet.html

- Identifying searching capabilities in search engines

In-class exercise reviewing & reporting on search engines
Identify the nature of information — what kinds of things have you found?

- Reports vs. proceedings vs. journal articles vs. books vs. reference (using reference as background for indiv topic or overview to recommend)

Intro to Voyager, the online catalog...
Homework Two Due: Revised Topic For Research

SESSION SIX

Now that I found useful information, what do I do?

- Simple searching in Voyager—background info

Intro: Create a working bibliography using the MLA format

Develop a portfolio or report (creating Web page: Word, Composer, HTML tags)
More on searching for books...

SESSION SEVEN

How do I really know if my information is good enough?
Identify the elements that determine appropriateness or authority of a source
Evaluate the quality of the information
Homework Three Due: Background Sources for Research

SESSION EIGHT

Information overload!!! Enough is enough—how do I organize it?
- Recognize the purpose and structure of disseminating information
- Organize information into a Web-based portfolio

SESSION NINE

I have a lot of information. How will I use it effectively?
Revise the information retrieval strategies as needed to improve the research
Homework Four Due: In-depth Sources for Research

SESSION TEN

Am I going in the right direction?
- In-class review of progress

SESSION ELEVEN

More searching?
- Use help pages to identify advanced searching techniques
- Formulate searches using Boolean and proximity operators
- Compare results using basic vs. advanced searching techniques

Homework Five Due: Up-to-date Sources for Research

SESSION TWELVE

Am I there yet?
- Identify relevance or precision of search results
- Increase relevance/precision of results using advanced techniques

SESSION THIRTEEN

What about information specific to my field? Like—patents?

- Describe the difference between specialized and general resources
- Find three types of specialized resources

SESSION FOURTEEN

So, how does this course really affect my life?

- Identify three key sources from your final project
- Extrapolate two things learned while searching
- Create an outline for the presentation

SESSION FIFTEEN

Presentations

SESSION SIXTEEN

Presentations

HOMEWORK #1

Each person should turn in the following information:

√ A problem statement for the individual research problem
(What product or enhancement are you going to investigate?)

√ Conditions to support your problem statement
(Who would benefit from this? How? Where? When?)

√ An hypothesis for this research problem
(What will your research prove or disprove?)

HOMEWORK #2

Each person should turn in the following information:

√ Revised problem statement for the individual research problem
(What product or enhancement are you going to investigate?)

√ Additional or new conditions to support your problem statement
(Who would benefit from this? How? Where? When?)

√ Revised hypothesis for this research problem
(What will your research prove or disprove?)

HOMEWORK #3

Each person should turn in the following information:

√ Photocopy of title page of book to be used as background resource
(You may need the reverse of the title for additional information)

- √ Source information for the book written out in proper citation format (http://gemini.lib.purdue.edu/instruction/gs175/Spring99/1gs175e/cite.html)
- √ An annotation describing how this book supports your topic *(how would it help someone who doesn't know this topic?)*

HOMEWORK #4

Each person should turn in the following information:

- √ Copy of specific journal article to be used as in-depth resource (you may need the reverse of the title for additional information)
- √ Source information for the journal article in proper citation format (http://gemini.lib.purdue.edu/instruction/gs175/Spring99/1gs175e/cite.html)
- √ An annotation describing how this journal article supports your topic *(how would it help someone who doesn't know this topic?)*

HOMEWORK #5

Each person should turn in the following information:

- √ Copy of specific Web page to be used as current resource *(you may need the reverse of the title for additional information)*
- √ Source (citation) information for the Web page in proper citation format (http://gemini.lib.purdue.edu/instruction/gs175/Spring99/1gs175c/cite.html)
- √ An annotation describing how this Web page supports your topic *(how would it help someone who doesn't know this topic?)*

INFORMATION STRATEGIES FOR PRODUCT ANALYSIS (FINAL PROJECT)

The outcome of this course is for students to use knowledge of information strategies to analyze technology products to produce a Final Project. This product analysis is a report based on the following scenario, and must include the elements listed below.
Scenario: You are working for a company which is interested in increasing its market share by extending its product line. Your job is to investigate a technology product or service and analyze it to make a recommendation on whether the company should

invest in developing a rival, but possibly more enhanced, product.

Introduction: What is the product or service? What are its uses? (By whom? For what? Where?) Why did you choose this one? *["Tell 'em what you're going to tell 'em."]*

Background: Overview of the general technology topic. (Assume the person for whom you are writing the report is not familiar with the technology.) Provide information about the product, and include reference to resources (books, encyclopedia articles, etc.) that the reader of your report will find useful to learn some background information.

Analysis:

- Consumer Market: Strengths and weaknesses; what it does well (or doesn't); who it is aimed at or designed for; current availability, or market standing
- Research & Development: Trends in current research for this product or service; possible future outcomes based on research; insights into how the company might do things differently in manufacture/marketing
- Specifications: Standards which it meets; technical specifications; related patents; other technical considerations

Conclusion: Summary of your work ["Tell 'em what you told 'em."] Include here descriptions of any problems you had in searching which may have affected the quality of information you found. Compare with other product or services you encountered.

Recommendation: This is where you make the case as to how the company should act on your product analysis. Give your opinion based on the research you've done. Should the company offer this product or service? Should they work from scratch? Create a similar product with enhancements? (What are they?) Should they just try to "buy the rights" of a currently available product? Sell your pitch!

The product analysis ***must be backed up with references***! You must have a total of ten references, including at least one from each of the following categories:

- Book
- Encyclopedia or Handbook
- Journal article
- Patent or Standard
- Web page or Product catalog

INDEX

M

O

P

Q

R

S

T

V

W

ABOUT THE AUTHOR

D. SCOTT BRANDT is a Professor of Library Science and the Technology Training Librarian at the Purdue University Libraries. He has been in that position since 1993. Working in a team environment, his responsibilities include: planning, organizing, coordinating, and conducting a training and instruction program for library staff, students, and faculty in the use of information technologies.

As a researcher, Brandt is interested in the mental models employed when using and learning about information technology. In 2001 he served as a Research Fellow at the University of Staffordshire, U.K., working on a project called "Applying Cognitive Task Analysis to Information Technology Learning." He is also working in the area of "Reusable Learning Objects," which is a natural outgrowth of the instructional systems design concepts used in this book.

Brandt has been a presenter and speaker on the topic of innovation in training since 1991. He gives several workshops yearly, nationally and internationally, in the use of applying educational theory to developing training in libraries. He is the author of several journal articles, and writes a column for Information Today's *Computers In Libraries* called "techman's techpage." He has been a conference program committee member for both Internet Librarian and Computers In Libraries since 1998.

In addition to helping people understand the concepts behind technology, he enjoys traveling, cooking, the ocean, and jazz and blues music.